Springer Series in Information Sciences 14

Editor: King-sun Fu

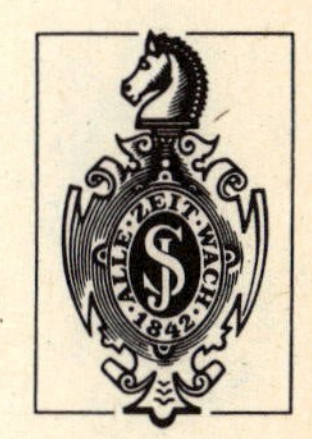

Springer Series in Information Sciences

Editors: King-sun Fu   Thomas S. Huang   Manfred R. Schroeder

---

Volume 1 **Content-Addressable Memories** By T. Kohonen

Volume 2 **Fast Fourier Transform and Convolution Algorithms** By H. J. Nussbaumer 2nd Edition

Volume 3 **Pitch Determination of Speech Signals** Algorithms and Devices By W. Hess

Volume 4 **Pattern Analysis** By H. Niemann

Volume 5 **Image Sequence Analysis** Editor: T. S. Huang

Volume 6 **Picture Engineering** Editors: King-sun Fu and T. L. Kunii

Volume 7 **Number Theory in Science and Communication** With Applications in Cryptography, Physics, Biology, Digital Information, and Computing. By M. R. Schroeder

Volume 8 **Self-Organization and Associative Memory** By T. Kohonen

Volume 9 **An Introduction to Digital Picture Processing** By L. P. Yaroslavsky

Volume 10 **Probability, Statistical Optics, and Data Testing** A Problem Solving Approach. By B. R. Frieden

Volume 11 **Physical and Biological Processing of Images** Editors: O. J. Braddick and A. C. Sleigh

Volume 12 **Multiresolution Image Processing and Analysis** Editor: A. Rosenfeld

Volume 13 **VLSI for Pattern Recognition and Image Processing** Editor: King-sun Fu

Volume 14 **Mathematics of Kalman-Bucy Filtering** By P. A. Ruymgaart and T. T. Soong

Volume 15 **The Formalization of Pattern Recognition Bases** By M. Pavel

Volume 16 **Introduction to Statistical Radiophysics and Optics I** Random Oscillations and Waves By S. A. Akhmanov, Y. Y. Dyakov, and A. S. Chirkin

---

P. A. Ruymgaart T. T. Soong

Mathematics of Kalman-Bucy Filtering

With 19 Figures

Springer-Verlag
Berlin Heidelberg New York Tokyo 1985

Dr. Peter A. Ruymgaart

Department of Mathematics, University of Technology, Delft,
Delft, The Netherlands

Professor Tsu T. Soong

Faculty of Engineering and Applied Sciences, State University of New York at Buffalo,
Buffalo, NY 14260, USA

ISBN 3-540-13508-1 Springer-Verlag Berlin Heidelberg New York Tokyo
ISBN 0-387-13508-1 Springer-Verlag New York Heidelberg Berlin Tokyo

Library of Congress Cataloging in Publication Data. Ruymgaart, P. A. (Peter Arnold), 1925-. Mathematics of Kalman-Bucy filtering. (Springer series in information sciences ; v. 14). Includes bibliographies and index. 1. Kalman filtering. 2. Probabilities. 3. Calculus. 4. Hilbert space. I. Soong, T. T. II. Title. III. Series. QA402.3.R89 1985 519.2 84-20293

Typesetting: Schwetzinger Verlagsdruckerei, 6830 Schwetzingen
Offset printing: Beltz Offsetdruck, 6944 Hemsbach/Bergstr.
Bookbinding: J. Schäffer OHG, 6718 Grünstadt
2153/3130-543210

to Alma, Tiemo, Anton

and

to Dottie, Karen, Stephen and Susan

Preface

Since their introduction in the mid 1950s, the filtering techniques developed by Kalman, and by Kalman and Bucy have been widely known and widely used in all areas of applied sciences. Starting with applications in aerospace engineering, their impact has been felt not only in all areas of engineering but also in the social sciences, biological sciences, medical sciences, as well as all other physical sciences. Despite all the good that has come out of this development, however, there have been misuses because the theory has been used mainly as a tool or a procedure by many applied workers without them fully understanding its underlying mathematical workings.

This book addresses a mathematical approach to Kalman-Bucy filtering and is an outgrowth of lectures given at our institutions since 1971 in a sequence of courses devoted to Kalman-Bucy filters. The material is meant to be a theoretical complement to courses dealing with applications and is designed for students who are well versed in the techniques of Kalman-Bucy filtering but who are also interested in the mathematics on which these may be based.

The main topic addressed in this book is continuous-time Kalman-Bucy filtering. Although the discrete-time Kalman filter results were obtained first, the continuous-time results are important when dealing with systems developing in time continuously, which are hence more appropriately modeled by differential equations than by difference equations. On the other hand, observations from the former can be obtained in a discrete fashion. Since it is shown that the Kalman estimator depends strongly on the mathematical structure of the observations, the question arises as to what degree this estimator depends on the mesh of time intervals between observations. The answer is contained in the mathematical model by replacing discrete observations by those continuous in time. This process leads to the Kalman-Bucy filter, the continuous version of the discrete Kalman filter.

In confining our attention to Kalman-Bucy filtering, the mathematics needed consists mainly of operations in Hilbert spaces. Thus, the mathematical development presented in this book is accomplished without the need for Ito calculus, or for the theory of martingales and Markov processes, or for the more artificial infinite-dimensional innovations approach.

The necessary basic concepts of probability theory and Hilbert spaces are reviewed in Chap. 1. A relatively complete treatment of mean-square calculus is given in Chap. 2, leading to the discussion of the Wiener-Lévy process. Chapter 3 treats the stochastic differential equations central to modeling the Kalman-Bucy filtering process.

A mathematical theory of the Kalman-Bucy filter is introduced in Chap. 4. Representation of the Kalman-Bucy estimator as an integral is presented, which is seen to lead in a natural way to the introduction of another type of integrals. This is the first opportunity to draw attention to the paradoxical situation in which the success of the filter depends on the noise in the observations. This is followed by the derivation of the Wiener-Hopf equation for the filter and by the computations leading to the Riccati equation and the ordinary differential equation satisfied by the Kalman-Bucy estimator.

With the aid of a theorem by Liptser and Shiryayev, Chap. 5 is intended to shed some light on the dependence of the Kalman-Bucy estimator on observation noise.

Some exercises are included to extend or amplify the text material. Solutions to some of these are supplied in the Appendix.

We wish to acknowledge support of the U.S. National Science Foundation. Through its fellowship and research grant awards we were able to meet and to begin this joint writing project. The first author is indebted to the late Prof. Dr. R. Timman for his encouragement.

We are also grateful to Mrs. Ikuko Isihara who patiently and expertly typed several versions of this manuscript.

Delft, The Netherlands and
Buffalo, New York, July, 1984

P. A. Ruymgaart
T. T. Soong

Contents

1. Elements of Probability Theory 1
1.1 Probability and Probability Spaces 1
1.1.1 Measurable Spaces, Measurable Mappings and Measure Spaces 1
1.1.2 Probability Spaces 3
1.2 Random Variables and "Almost Sure" Properties 3
1.2.1 Mathematical Expectations 5
1.2.2 Probability Distribution and Density Functions 7
1.2.3 Characteristic Function 9
1.2.4 Examples 9
1.3 Random Vectors 11
1.3.1 Stochastic Independence 13
1.3.2 The Gaussian N Vector and Gaussian Manifolds 16
1.4 Stochastic Processes 18
1.4.1 The Hilbert Space $L_2(\Omega)$ 21
1.4.2 Second-Order Processes 24
1.4.3 The Gaussian Process 26
1.4.4 Brownian Motion, the Wiener-Lévy Process and White Noise 27

2. Calculus in Mean Square 30
2.1 Convergence in Mean Square 30
2.2 Continuity in Mean Square 33
2.3 Differentiability in Mean Square 35
2.3.1 Supplementary Exercises 39
2.4 Integration in Mean Square 40
2.4.1 Some Elementary Properties 42
2.4.2 A Condition for Existence 46
2.4.3 A Strong Condition for Existence 52
2.4.4 A Weak Condition for Existence 54
2.4.5 Supplementary Exercises 63
2.5 Mean-Square Calculus of Random N Vectors 65
2.5.1 Conditions for Existence 67

2.6 The Wiener-Lévy Process 68
2.6.1 The General Wiener-Lévy N Vector 70
2.6.2 Supplementary Exercises 75
2.7 Mean-Square Calculus and Gaussian Distributions 76
2.8 Mean-Square Calculus and Sample Calculus 76
2.8.1 Supplementary Exercise 79

3. The Stochastic Dynamic System 80
3.1 System Description 80
3.2 Uniqueness and Existence of m.s. Solution to (3.3) 81
3.2.1 The Banach Space $L_2^N(\Omega)$ 81
3.2.2 Uniqueness 83
3.2.3 The Homogeneous System 84
3.2.4 The Inhomogeneous System 89
3.2.5 Supplementary Exercises 93
3.3 A Discussion of System Representation 94

4. The Kalman-Bucy Filter 100
4.1 Some Preliminaries 101
4.1.1 Supplementary Exercise 102
4.2 Some Aspects of $L_2([a, b])$ 103
4.2.1 Supplementary Exercise 106
4.3 Mean-Square Integrals Continued 106
4.4 Least-Squares Approximation in Euclidean Space 114
4.4.1 Supplementary Exercises 116
4.5 A Representation of Elements of $H(\mathbf{Z}, t)$ 116
4.5.1 Supplementary Exercises 128
4.6 The Wiener-Hopf Equation 128
4.6.1 The Integral Equation (4.106) 134
4.7 Kalman-Bucy Filter and the Riccati Equation 146
4.7.1 Recursion Formula and the Riccati Equation 148
4.7.2 Supplementary Exercise 153

5. A Theorem by Liptser and Shiryayev 154
5.1 Discussion on Observation Noise 154
5.2 A Theorem of Liptser and Shiryayev 155

Appendix: Solutions to Selected Exercises 158

References 167

Subject Index 169

1. Elements of Probability Theory

As the Kalman-Bucy filtering theory is a probabilistic concept, an understanding of some basic concepts in probability theory is necessary in the study of this subject. We begin this discourse by reviewing some of the basic elements in probability theory. Details and proofs can be found in [1.1–4], for example.

1.1 Probability and Probability Spaces

The mathematical theory of probability gives us the basic tools for constructing and analyzing random phenomena. In studying a random phenomenon, we are dealing with an experiment whose outcome is not predictable with certainty, i.e., a *random experiment*.

A typical random experiment is that of throwing a die; the possible outcomes are known to be 1, 2, 3, 4, 5 and 6, but it is not known in advance which number will come up in any trial. The set of all possible outcomes of a random experiment is called the *sample space* and subsets of the sample space are called *events*. In the die-throwing experiment, for example, the sets $\{2, 4, 6\}$ (even numbers), $\{6\}$ (the number six), and $\{1, 2, 3, 4, 5\}$ (not six) are events. Our interest in the study of a random experiment is in the statements we can make concerning the events that can occur; these statements are expressed as *probabilities* of events. In the development of the probability measure or probability, we first consider some fundamental notions in measure theory.

1.1.1 Measurable Spaces, Measurable Mappings and Measure Spaces

Let Ω be a nonempty set with subsets $A, B, C, \dots$. Some of the standard set operations together with their usual notations are given below.

Union: $A \cup B$
Intersection: $A \cap B$
Complement of A: $A^{\mathrm{C}} = \Omega \setminus A$
Empty set: $\emptyset$.

A set $\mathscr{A}$ of subsets of Ω is a *σ algebra* if
a) The empty set $\emptyset$ belongs to $\mathscr{A}$.
b) If $A \in \mathscr{A}$, then $A^C \in \mathscr{A}$.
c) If $A_i \in \mathscr{A}, \quad i \in \mathbb{N}, \quad$ then $\bigcup_{i=1}^{\infty} A_i \in \mathscr{A}$.

It then follows that $\Omega \in \mathscr{A}$ and that $\mathscr{A}$ also contains the unions and intersections of all finite and denumerable collections of its elements.

The elements of $\mathscr{A}$ are called *measurable sets* and the pair $\{\Omega, \mathscr{A}\}$ is called a *measurable space*. Given Ω, there are in general many different σ algebras and measurable spaces, the largest σ algebra containing all subsets of Ω, while the smallest contains only Ω and $\emptyset$ as elements.

Let $\{\Omega_1, \mathscr{A}_1\}$ and $\{\Omega_2, \mathscr{A}_2\}$ be measurable spaces. The mapping $\Phi : \Omega_1 \to \Omega_2$ is called measurable if for each $A \in \mathscr{A}_2$, $\Phi^{-1}A \in \mathscr{A}_1$, $\Phi^{-1}A$ being $\{\omega \in \Omega_1 : \Phi(\omega) \in A\}$.

If $\{\Omega_3, \mathscr{A}_3\}$ is also a measurable space and $\Psi : \Omega_2 \to \Omega_3$ a measurable mapping, it is seen that the composed mapping

$$\Psi \cdot \Phi : \Omega_1 \to \Omega_3$$

is also measurable. Here, the measurable spaces used may be identical.

Consider the mapping $\mu : \mathscr{A} \to \mathbb{R}$. It is a *measure* if it satisfies the following conditions:
1) *$\mu A \geqslant 0$ for all $A \in \mathscr{A}$;*
2) there is at least one element $F \in \mathscr{A}$ such that μF is finite;
3) if $(A_i)_{i \in \mathbb{N}}$ is a sequency of mutually disjoint elements of $\mathscr{A}$, then for each $n \in \mathbb{N}$,

$$\mu \bigcup_{i=1}^{n} A_i = \sum_{i=1}^{n} \mu A_i \quad \text{(additivity)}$$

$$\mu \bigcup_{i=1}^{\infty} A_i = \lim_{n \to \infty} \sum_{i=1}^{n} \mu A_i \quad (\sigma \text{ additivity}) .$$

A number of other properties can be derived as a result of these conditions. For example, we have $\mu\emptyset = 0$.

The triple $\{\Omega, \mathscr{A}, \mu\}$ is called a *measure space*.

Suppose that Ω is $\mathbb{R}^n$, or some interval $I_1 \times I_2 \times \ldots \times I_n$ of $\mathbb{R}^n$, where I_k, $k = 1, \ldots, n$, is an interval of $\mathbb{R}$, $n \in \mathbb{N}$. The *Borel algebra* $\mathscr{B}^n$ is defined as the σ algebra generated by the intervals of this Ω. This means that $\mathscr{B}^n$ is the intersection of all σ algebras containing the intervals of $\mathbb{R}^n$, where the intervals can be arbitrarily specified (open, closed, or arbitrary). Let us note that the σ algebra consisting of all subsets of $\mathbb{R}^n$ is not the Borel algebra $\mathscr{B}^n$, although it is not easy to define a subset of $\mathbb{R}^n$ not belonging to $\mathscr{B}^n$.

If $m \in \mathbb{N}$ also, a measurable mapping from $\{\mathbb{R}^m, \mathscr{B}^m\}$ into $\{\mathbb{R}^n, \mathscr{B}^n\}$ is called a *Borel function*. Hence, a mapping $f : \mathbb{R}^m \to \mathbb{R}^n$ is a Borel function if and only if for each $B \in \mathscr{B}^n$, $f^{-1}B \in \mathscr{B}^m$. It follows that the class of Borel functions mapping $\mathbb{R}^m$ into $\mathbb{R}^n$ contains the class of continuous mappings from $\mathbb{R}^m$ into $\mathbb{R}^n$.

There is just one measure λ^n on $\mathscr{B}^n$ with the property that for every interval $[a_1, b_1] \times \ldots \times [a_n, b_n]$ the measure is equal to $(b_1 - a_1)(b_2 - a_2) \ldots (b_n - a_n)$. This measure is called the *Lebesgue measure*.

The triple $\{\mathbb{R}^n, \mathscr{B}^n, \lambda^n\}$ defines a *Lebesgue measure space*. If $n = 1$, one simply writes $\{\mathbb{R}, \mathscr{B}, \lambda\}$ or $\{I, \mathscr{B}, \lambda\}$, where I is some interval of $\mathbb{R}$.

1.1.2 Probability Spaces

A *probability space* $\{\Omega, \mathscr{A}, P\}$ is a measure space to which the measure P assigns the value 1 to Ω. Hence,

$$P : \mathscr{A} \rightarrow [0, 1] .$$

The measure P is called the *probability measure* or simply *probability*.

In applications, Ω is an abstraction of the sample space and the elements of $\mathscr{A}$ are abstractions of events. In the die-throwing experiment discussed in Sect. 1.1, for example, a suitable probability space may be specified as follows:

$$\Omega = \{\omega_i, i = 1, 2, \ldots, 6\} ,$$

$$\mathscr{A} = \text{class of all subsets of } \Omega ,$$

$$P\{\omega_i\} = 1/6 , i = 1, \ldots, 6 .$$

An alternative is

$$\Omega = (0, 6] ,$$

$$\mathscr{A} = \text{the Borel algebra } \mathscr{B} \text{ of } (0, 6] ,$$

$$\forall B \in \mathscr{B}, PB = \text{number of integers belonging to } B \text{ divided by integer } 6 .$$

Only the empty set has probability zero in the first case, whereas in the second case the probability of each $B \in \mathscr{B}$ containing no integers is equal to zero.

1.2 Random Variables and "Almost Sure" Properties

Let $\mathbb{R}$ and $\mathscr{B}$ be as defined in Sect. 1.1.1 and $\{\Omega, \mathscr{A}, P\}$ as in Sect. 1.1.2. A measurable mapping

$$X : \Omega \to \mathbb{R} \tag{1.1}$$

is called a *random variable* (r.v.).

As composed measurable functions are measurable, it follows that the composition

$$f \cdot X$$

of a r.v. X and a Borel function f is a random variable. For example, $|X|$ and X^2 are random variables if X is a random variable.

A random variable X is called *simple* if its range $X\Omega \subset \mathbb{R}$ is a finite set, say,

$$X\Omega = \{a_1, \ldots, a_n\} \subset \mathbb{R} . \tag{1.2}$$

For any subset $S \subset \Omega$, the indicator function $i_S : \Omega \to \mathbb{R}$ is defined by

$$i_S(\omega) = \begin{cases} 1 & \text{if} \quad \omega \in S \\ 0 & \text{if} \quad \omega \notin S . \end{cases} \tag{1.3}$$

Hence, if X is the simple r.v. given in (1.2) and if

$$X^{-1}\{a_j\} = A_j , \quad j = 1, \ldots, n, \quad A_j \in \mathscr{A} , \tag{1.4}$$

it is seen that

$$X = \sum_{j=1}^{n} a_j i_{A_j} . \tag{1.5}$$

It can be shown that a mapping $X : \Omega \to \mathbb{R}$ is a random variable if and only if it is the pointwise limit of a sequence of simple random variables. Hence, if X and Y are r.v.'s and if $c \in \mathbb{R}$, the mappings cX, $|X|$, X^2, $X + Y$, XY, etc., are also random variables. We note that the set of all random variables defined on Ω is a vector space formed under usual addition and multiplication with a scalar.

Let us now define an "almost sure" property associated with two random variables. Let r.v.'s X and Y be defined on Ω. They are called *almost sure identical* with the notation $X = Y$ a.s. if $X(\omega) = Y(\omega)$ for all $\omega \in \Omega$ outside $N \in \mathscr{A}$ with $PN = 0$ or, equivalently, $X(\omega) = Y(\omega)$ for all $\omega \in \Omega'$, where $\Omega' \in \mathscr{A}$ and $P\Omega' = 1$.

If $X = Y$ a.s., it will be shown in the following sections that they have the same probabilistic properties. Since

$$\text{“}X \approx Y \quad \text{if} \quad X = Y \text{ a.s.”}$$

is an equivalence relation, the set of all random variables defined on $\{\Omega, \mathscr{A}, P\}$ may be partitioned into disjoint equivalence classes. From the probabilistic point of view, the elements of a class are identical. We shall also see that since it will not always be possible to specify a random variable on all of Ω, but only on some measurable set Ω' with $P\Omega' = 1$, it is more expedient to work with equivalence classes of random variables than with the individual random variables themselves. Quite often we shall write "*a* r.v. X" when an equivalence class is meant.

Let us now consider two sets of r.v.'s $\{X_n : \Omega \to \mathbb{R},\ n \in \mathbb{N}\}$ and $\{X_n' : \Omega \to \mathbb{R}, n \in \mathbb{N}\}$, where $X_n = X_n'$ a.s. for each $n \in \mathbb{N}$. This means that at each $n \in \mathbb{N}$, there is a set $N_n \in \mathscr{A}$ with $PN_n = 0$ such that

$$X_n(\omega) = X_n'(\omega)$$

if $\omega \in \Omega \backslash N_n$. Then

$$\forall n \in \mathbb{N}: X_n(\omega) = X_n'(\omega) \quad \text{if} \quad \omega \in \Omega' = \Omega \backslash \bigcup_{n=1}^{\infty} N_n \,.$$

It is easily shown that

$$P\bigcup_{n=1}^{\infty} N_n \leqslant \sum_{n=1}^{\infty} PN_n = 0$$

and hence

$$P\Omega' = 1.$$

Thus the systems $\{X_n,\ n \in \mathbb{N}\}$ and $\{X_n',\ n \in \mathbb{N}\}$ are identical on a set of probability one. Now, if $X : \Omega \to \mathbb{R}$ is a random variable and if X_n tends to X pointwise on Ω, then at each $\omega \in \Omega'$ we also have

$$\lim_{n\to\infty} X_n'(\omega) = X(\omega)$$

and we write "$X' \to X$ a.s.".

However, certain difficulties may arise if the number of random variables in a system is no longer denumerable. An example of this situation is given in Sect. 1.4 and so is discussed briefly there.

Below, all random variables considered in a given situation are understood to be defined on one and the same probability space unless stated otherwise.

1.2.1 Mathematical Expectations

Let $\{\Omega, \mathscr{A}, P\}$ be a probability space and let X be a simple random variable defined by (1.5). Its *mathematical expectation* or *mean*, $\mathrm{E}X$, is defined as

$$\mathrm{E}X = \sum_{i=1}^{n} a_i PA_i \,. \tag{1.6}$$

The notation

$$\mathrm{E}X = \int_{\Omega} X(\omega) P(d\omega) \quad \text{or} \tag{1.7}$$

$$\mathrm{E}X = \int_{\Omega} X(\omega)\, dP(\omega) \tag{1.8}$$

is also used sometimes.

If X and Y are simple random variables, then

$$X = Y \text{ a.s.} \Rightarrow \mathrm{E}X = \mathrm{E}Y \quad \text{and} \tag{1.9}$$

$$X \leqslant Y \text{ a.s.} \Rightarrow \mathrm{E}X \leqslant \mathrm{E}Y\,. \tag{1.10}$$

Consider a r.v. X and define the r.v.'s X^+ and X^- as follows:

$$X^+(\omega) = \begin{cases} X(\omega) & \text{if } X(\omega) \geqslant 0 \\ 0 & \text{if } X(\omega) < 0 \end{cases} \tag{1.11}$$

and

$$X^-(\omega) = \begin{cases} 0 & \text{if } X(\omega) \geqslant 0 \\ -X(\omega) & \text{if } X(\omega) < 0\,. \end{cases} \tag{1.12}$$

Then

$$X^+ \geqslant 0\,, \quad X^- \geqslant 0 \quad \text{and} \tag{1.13}$$

$$X = X^+ - X^-\,, \quad |X| = X^+ + X^-\,. \tag{1.14}$$

It can be shown that nonnegative random variables may be seen as limits of nondecreasing sequences of simple random variables. Hence there are sequences $\{X_n^+\}_{n\in\mathbb{N}}$ and $\{X_n^-\}_{n\in\mathbb{N}}$ of simple random variables such that

$$X_n^+ \uparrow X^+ \quad \text{and} \quad X_n^- \uparrow X^- \quad \text{as} \quad n \to \infty\,.$$

Likewise, the sequences

$$\{\mathrm{E}X_n^+\}_{n\in\mathbb{N}} \quad \text{and} \quad \{\mathrm{E}X_n^-\}_{n\in\mathbb{N}}$$

are monotone. If they do not tend to infinity, they converge to finite limits. We then define

$$\mathrm{E}X^+ = \lim_{n\to\infty} \mathrm{E}X_n^+\,, \quad \mathrm{E}X^- = \lim_{n\to\infty} \mathrm{E}X_n^- \quad \text{and}$$

$$\mathrm{E}X = \mathrm{E}X^+ - \mathrm{E}X^-\,.$$

These definitions are admissible in the sense that different monotone sequences tending to X^+ (or X^-) lead to the same limit $\mathrm{E}X^+$ (or $\mathrm{E}X^-$).

Let us now note some properties of the mathematical expectation "operator" E. Consider r.v.'s X and Y with finite expectations and let $c \in \mathbb{R}$. We can easily verify that

$$\begin{aligned} &\mathrm{E}\{cX\} = c\mathrm{E}X \\ &\mathrm{E}\{X+Y\} = \mathrm{E}X + \mathrm{E}Y \\ &X \leqslant Y \Rightarrow \mathrm{E}X \leqslant \mathrm{E}Y . \end{aligned} \tag{1.15}$$

Since $|X| = X^+ + X^-$, we also have

$$\begin{aligned} &\mathrm{E}|X| = \mathrm{E}X^+ + \mathrm{E}X^- \\ &|\mathrm{E}X| \leqslant \mathrm{E}|X| . \end{aligned} \tag{1.16}$$

If $A \in \mathscr{A}$ and if i_A is the indicator function of A, then

$$\mathrm{E}\, i_A = PA \tag{1.17}$$

and, in particular,

$$\mathrm{E}\, i_\Omega = P\Omega = 1 . \tag{1.18}$$

A random variable which is almost sure equal to a given real number x is called *degenerate*. It can be represented by xi_Ω and

$$\mathrm{E}\{xi_\Omega\} = x\,\mathrm{E}\, i_\Omega = x . \tag{1.19}$$

Conversely, every constant may be seen as a degenerate random variable. In particular, $\mathrm{E}X$ may be seen as $(\mathrm{E}X)\, i_\Omega$. Hence, we can write

$$\mathrm{E}\{\mathrm{E}X\} = \mathrm{E}\{(\mathrm{E}X)\, i_\Omega\} = \mathrm{E}X \quad \text{and} \tag{1.20}$$

$$\mathrm{E}\{X - \mathrm{E}X\} = \mathrm{E}X - \mathrm{E}\{\mathrm{E}X\} = \mathrm{E}X - \mathrm{E}X = 0 \tag{1.21}$$

provided that $\mathrm{E}X$ is finite. Equation (1.21) shows that r.v. $X - \mathrm{E}X$ has zero mathematical expectation or zero mean. Random variables with zero means are called *centered* (at expectation). If $\mathrm{E}\{X - \mathrm{E}X\}^2$ exists, it is called the *variance* of X.

1.2.2 Probability Distribution and Density Functions

The *probability distribution function* F_X of a r.v. X is defined by

$$F_X(x) = P\{\omega : X(\omega) < x\} , \quad x \in \mathbb{R} . \tag{1.22}$$

It exists for all random variables and

a) it is nondecreasing

b) it is continuous from the left

c) $\lim_{x\to\infty} F_X(x) = 1$ and $\lim_{x\to-\infty} F_X(x) = 0$.

We also note that, conversely, any function $F: \mathbb{R} \to [0, 1]$ with the aforementioned properties is a probability distribution function of some random variable. Indeed, given F, there are always probability spaces $\{\Omega, \mathscr{A}, P\}$ and measurable mappings $X: \Omega \to \mathbb{R}$ whose probability distribution function F_X is equal to F. Furthermore, as we anticipated in Sect. 1.2, all elements of the equivalence class of X have the same probability distribution function F, i.e.,

$$X = Y \text{ a.s.} \Rightarrow F_X = F_Y . \tag{1.23}$$

From the definition (1.22), we see that, given two real numbers a and b with $b > a$,

$$P\{\omega : a \leqslant X(\omega) < b\} = F_X(b) - F_X(a) . \tag{1.24}$$

It can be shown that

$$\mathrm{E}X = \int_{-\infty}^{\infty} x \, dF_X(x) \tag{1.25}$$

and, if $g: \mathbb{R} \to \mathbb{R}$ is a Borel function,

$$\mathrm{E}\{g \cdot X\} = \int_{\Omega} g[X(\omega)] \, dP(\omega) = \int_{-\infty}^{\infty} g(x) \, dF_X(x) \tag{1.26}$$

where the integrals in (1.25, 26) are of Riemann-Stieltjes type. In particular,

$$\mathrm{E}X^2 = \int_{-\infty}^{\infty} x^2 \, dF_X(x) . \tag{1.27}$$

When there is a continuous mapping $f_X: \mathbb{R} \to \mathbb{R}$ such that

$$F_X(x) = \int_{-\infty}^{x} f_X(u) \, du ,$$

the mapping f_X is called the *probability density function* of r.v. X. In terms of $f_X(x)$, we can verify that

$$P\{\omega : a \leqslant X(\omega) < b\} = \int_{a}^{b} f_X(x) \, dx \tag{1.28}$$

$$\mathrm{E}X = \int_{-\infty}^{\infty} x f_X(x) \, dx \quad \text{and} \tag{1.29}$$

$$\mathrm{E}\{g \cdot X\} = \int_{-\infty}^{\infty} g(x) f_X(x) \, dx , \tag{1.30}$$

the integrals being Riemann type.

1.2.3 Characteristic Function

Associated with a probability distribution function F_X, there exists a *characteristic function* ϕ_X defined by the Riemann-Stieltjes integral

$$\phi_X(u) = \mathrm{E}\{e^{iuX}\} = \int_{-\infty}^{\infty} e^{iux}\, dF_X(x)\,, \quad u \in \mathbb{R}\,, \tag{1.31}$$

where $i^2 = -1$.

Given a characteristic function ϕ, it can be shown that there is a unique probability distribution function F with the properties (a), (b) and (c) of Sect. 1.2.2, whose characteristic function is ϕ. We remark that instead of "characteristic function of F_X" one can also write "characteristic function of X".

1.2.4 Examples

Let us now exemplify some of the basic concepts outlined in the preceding sections.

Example 1.1. Consider the probability space $\{\Omega, \mathscr{A}, P\}$ specified by

$$\Omega = \{\omega_1, \omega_2, \ldots, \omega_6\}$$

$$\mathscr{A} = \text{set of all subsets of } \Omega$$

$$P\{\omega_i\} = 1/6\,, \quad i = 1, 2, \ldots, 6\,.$$

Let $X : \Omega \to \mathbb{R}$ be defined by

$$X(\omega_i) = i\,, \quad i = 1, 2, \ldots, 6\,.$$

The mapping X is thus a simple random variable and may be used, for example, as a model for the die-throwing experiment; the probability that the outcome is i, $i = 1, 2, \ldots, 6$, is

$$P\{\omega : X(\omega) = i\} = P\{\omega_i\} = 1/6\,.$$

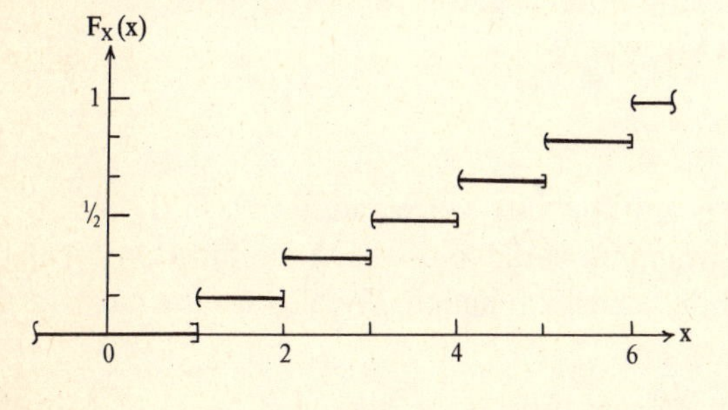

Fig. 1.1. $F_X(x) = P(X < x)$ in Example 1.1

The expectation of X is

$$EX = 1\left(\frac{1}{6}\right) + 2\left(\frac{1}{6}\right) + \ldots + 6\left(\frac{1}{6}\right) = 3.5 .$$

Similarly, the probability distribution function of X is

$$F_X(x) = \begin{cases} 0\,, & x \leqslant 1 \\ \frac{1}{6}\,, & 1 < x \leqslant 2 \\ \frac{2}{6}\,, & 2 < x \leqslant 3 \\ \frac{3}{6}\,, & 3 < x \leqslant 4 \\ \frac{4}{6}\,, & 4 < x \leqslant 5 \\ \frac{5}{6}\,, & 5 < x \leqslant 6 \\ 1\,, & 6 < x \end{cases}$$

and is presented graphically in Fig. 1.1. The characteristic function of X is

$$\phi_X(u) = \int_{-\infty}^{\infty} e^{iux}\, dF_X(x) = \frac{1}{6}\,[e^{iu} + e^{2iu} + \ldots + e^{6iu}]\,, \qquad u \in \mathbb{R} .$$

Example 1.2. Let m and $\sigma \in \mathbb{R}$ and $\sigma > 0$ and write

$$f_X(x) = \frac{1}{\sigma\sqrt{2\pi}} \exp\left[-\frac{1}{2}\left(\frac{x-m}{\sigma}\right)^2\right], \qquad x \in \mathbb{R}\,, \qquad \text{then} \tag{1.32}$$

$$F_X(x) = \int_{-\infty}^{x} f_X(u)\, du \tag{1.33}$$

is a probability distribution function. A r.v. X having this probability distribution function is called a *Gaussian* or *normal* random variable.

Consider the real line $\mathbb{R}$ with Borel algebra $\mathcal{B}$ and measure P on $\mathcal{B}$ induced by $F_X(x)$ given by (1.33), where

$$P[a, b) = F(b) - F(a) .$$

Then the resulting $\{\mathbb{R}, \mathcal{B}, P\}$ is a probability space and $X : \mathbb{R} \to \mathbb{R}$ defined by $X(x) = x$ for all $x \in \mathbb{R}$ is a random variable of which the function $f_X(x)$ defined by (1.32) is the probability density function. We also see the mean of X is

$$\mathrm{E}X = \int_{-\infty}^{\infty} x f_X(x)\, dx = m$$

and the variance, $\mathrm{E}(X-m)^2$, is

$$\mathrm{E}(X-m)^2 = \int_{-\infty}^{\infty} (x-m)^2 f_X(x)\, dx = \sigma^2 .$$

Thus, the two parameters m and σ^2 in $f_X(x)$ are, respectively, the mean and variance of r.v. X. Then

$$P\{|X-m| > \lambda\sigma\} = \sqrt{\frac{2}{\pi}} \int_{\lambda}^{\infty} \mathrm{e}^{-t^2/2}\, dt . \tag{1.34}$$

The characteristic function of X is

$$\phi_X(u) = \mathrm{E}\{\mathrm{e}^{\mathrm{i}uX}\} = \int_{-\infty}^{\infty} \mathrm{e}^{\mathrm{i}ux} f_X(x)\, dx = \exp\left(ium - \frac{u^2\sigma^2}{2}\right) . \tag{1.35}$$

1.3 Random Vectors

Consider now a system of n r.v.'s $X_1, X_2, \ldots, X_n$. It is convenient to represent them as components of an n-dimensional *random vector* $\boldsymbol{X}$ in the form

$$\boldsymbol{X} = \begin{bmatrix} X_1 \\ X_2 \\ \vdots \\ X_n \end{bmatrix} . \tag{1.36}$$

This is a measurable mapping of a propability space $\{\Omega, \mathcal{A}, P\}$ into the measurable space $\{\mathbb{R}^n, \mathcal{B}^n\}$. Hence, at each $\omega \in \Omega$, $\boldsymbol{X}(\omega)$ is an element of $\mathbb{R}^n$ and the components X_i, $i = 1, 2, \ldots, n$, are measurable mappings from $\{\Omega, \mathcal{A}, P\}$ into $\{\mathbb{R}, \mathcal{B}\}$.

The *joint probability distribution function* $F_{X_1X_2\ldots X_n}$ of $X_1, X_2, \ldots, X_n$ or $F_{\boldsymbol{X}} : \mathbb{R}^n \to [0, 1]$ of $\boldsymbol{X}$ is defined by

$$\begin{aligned} F_{\boldsymbol{X}}(\boldsymbol{x}) &= P\{\omega \in \Omega : X_i(\omega) < x_i, \quad i = 1, 2, \ldots, n\} \\ &= P\left[\bigcap_{i=1}^{n} \{\omega \in \Omega : X_i(\omega) < x_i\}\right] \end{aligned}$$

for each $\boldsymbol{x} \in \mathbb{R}^n$.

Defining

$$\begin{aligned} \underset{[a_i, b_i)}{\Delta} F(\boldsymbol{x}) = {} & F_{\boldsymbol{X}}(x_1, \ldots, x_{i-1}, b_i, x_{i+1}, \ldots, x_n) \\ & - F_{\boldsymbol{X}}(x_1, \ldots, x_{i-1}, a_i, x_{i+1}, \ldots, x_n) , \end{aligned} \tag{1.37}$$

where $a_i < b_i$, $i = 1, 2, \ldots, n$, we have the following properties for $F_X(x)$:

1) $$\underset{[a_n, b_n)}{\Delta}\left[\underset{[a_{n-1}, b_{n-1})}{\Delta}\left[\cdots\left[\underset{[a_1, b_1)}{\Delta} F(x)\right]\cdots\right]\right]$$
$$= P\{\omega \in \Omega : X(\omega) \in [a_1, b_1) \times \ldots [a_n, b_n)\} \geqslant 0 \tag{1.38}$$

2) $$F(x_1, \ldots, x_n) \to F(y_1, \ldots, y_n) \quad \text{as} \quad x_i \uparrow y_i\,, \quad i = 1, 2, \ldots, n \tag{1.39}$$

3) $F(\infty, \ldots, \infty) = 1$ and $F(x_1, \ldots, x_n) = 0$ if $x_i = -\infty$ for at least one $i \in \{1, \ldots, n\}$, where $F(\ldots, x_{i-1}, \infty\,(-\infty), \ldots)$ stands for $\lim\limits_{x_i \to \infty\,(-\infty)} F(\ldots, x_{i-1}, x_i, \ldots)$.

Conversely, a mapping $F : \mathbb{R}^n \to [0, 1]$ is a probability distribution function if it satisfies the properties stated above. Given F, there are always probability spaces $\{\Omega, \mathscr{A}, P\}$ and random n vectors $X : \Omega \to \mathbb{R}^n$ whose joint probability distribution functions F_X are equal to F. Let $\Omega = \mathbb{R}^n$ and $\mathscr{A} = \mathscr{B}^n$, for example. Then, if $P : \mathscr{B}^n \to [0, 1]$ is induced by F according to

$$P\{(-\infty, x_1) \times \ldots \times (-\infty, x_n)\} = F(x_1, \ldots, x_n)\,,$$

and if $X : \mathbb{R}^n \to \mathbb{R}^n$ is such that

$$X(x_1, \ldots, x_n) = \begin{bmatrix} x_1 \\ \vdots \\ x_n \end{bmatrix},$$

it is seen that F_X is identical to the given F.

In the context of n r.v.'s, joint probability distribution functions of subsets of $\{X_1, \ldots, X_n\}$ are called *marginal* distribution functions and are found from, for example,

$$F_{X_3}(x_3) = F_X(\infty, \infty, x_3, \infty, \ldots, \infty)$$

$$F_{X_1X_3}(x_1, x_3) = F_X(x_1, \infty, x_3, \infty, \ldots, \infty)\,.$$

When there is a continuous mapping $f_X : \mathbb{R}^n \to \mathbb{R}$ such that

$$F_X(x) = \int_{-\infty}^{x_n} \ldots \int_{-\infty}^{x_1} f_X(t_1, \ldots, t_n)\, dt_1 \ldots dt_n\,, \tag{1.40}$$

the function f_X is called the *joint probability density function* of X. In terms of F_X, it is given by

$$f_X(x) = \frac{\partial^n F_X(x)}{\partial x_1 \partial x_2 \ldots \partial x_n}\,. \tag{1.41}$$

As in the one-r.v. case, associated with a joint probability distribution function F_X there is a *joint characteristic function* given by

$$\phi_X(u) = \mathrm{E}\{\mathrm{e}^{\mathrm{i}u^T X}\}$$

$$= \int_{-\infty}^{\infty} \ldots \int_{-\infty}^{\infty} \mathrm{e}^{\mathrm{i}u^T x} d_1 \ldots d_n F_X(x)\,, \qquad u = \begin{bmatrix} u_1 \\ \vdots \\ u_n \end{bmatrix} \in \mathbb{R}^n\,. \tag{1.42}$$

There is a one-to-one correspondence between F and ϕ.

When the joint probability density function exists, we have

$$\phi_X(u) = \int_{-\infty}^{\infty} \ldots \int_{-\infty}^{\infty} \mathrm{e}^{\mathrm{i}u^T x} f_X(x)\, dx\,. \tag{1.43}$$

1.3.1 Stochastic Independence

The intuitive idea of *stochastic independence* or simply *independence* of events E_1 and E_2 is that the probability that E_1 and E_2 both occur is equal to the product of the probabilities of occurrences of individual events E_1 and E_2. Let us first give an example.

Example 1.3. Let $\{\Omega, \mathscr{B}^2, \lambda^2\}$ be a probability space with $\Omega = [0, 1] \times [0, 1]$, $\mathscr{B}^2$ its Borel algebra and λ^2 the Lebesgue measure. If I_1 and I_2 are intervals of $[0, 1]$ as shown in Fig. 1.2, we see that the sets $I_1 \times [0, 1]$ and $[0, 1] \times I_2$ are independent since

$$I_1 \times I_2 = \{I_1 \times [0, 1]\} \cap \{[0, 1] \times I_2\}$$

and thus

$$\begin{aligned} &P(\{I_1 \times [0, 1]\} \cap \{[0, 1] \times I_2\}) \\ &= P\{I_1 \times I_2\} = \lambda I_1 \times \lambda I_2 \\ &= P\{I_1 \times [0, 1]\} P\{[0, 1] \times I_2\} \end{aligned}$$

if P denotes λ^2.

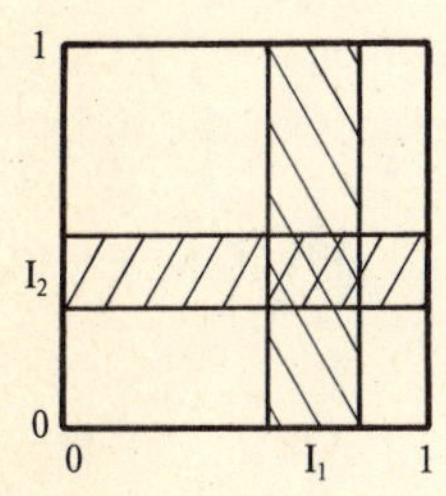

Fig. 1.2. The set $I_1 \times I_2$

Now, let $\{\Omega, \mathscr{A}, P\}$ be a probability space and let S be some set and $\mathscr{B}$ the Borel algebra of $\mathbb{R}$. By definition, a class

$$C = \{A_s, s \in S\} \subset \mathscr{A}$$

is independent if, for any finite subset $\{A_{s1}, \ldots, A_{sn}\} \subset C$, $\{s1, \ldots, sn\} \subset S$,

$$P\{A_{s1} \cap \ldots \cap A_{sn}\} = PA_{s1} PA_{s2} \ldots PA_{sn} . \tag{1.44}$$

For a set of classes $\{C_1, \ldots, C_n\}$, $C_i \subset \mathscr{A}$, $i = 1, 2, \ldots, n$, it is independent if, for all choices $\{i_1, \ldots, i_k\} \subset \{1, \ldots, n\}$ and for all choices $A_{i_j} \in C_{i_j}$, $j = 1, \ldots, k$, the class $\{A_{i_1}, \ldots, A_{i_k}\}$ is independent.

Consider now a set $\{X_s, s \in S\}$ of r.v.'s X_s defined on $\{\Omega, \mathscr{A}, P\}$. It is independent (or the random variables are independent) if, for any finite set $\{S_1, \ldots, S_n\} \subset S$, the set of classes

$$\{\{X_{sk}^{-1} B, B \in \mathscr{B}\}, \quad k = 1, \ldots, n\}$$

is independent.

Example 1.4. Let $\{\Omega, \mathscr{A}, P\} = \{[0, 1]^2, \mathscr{B}^2, \lambda^2\}$ as in Example 1.3 and let $\mathscr{B}$ be the Borel algebra of $[0, 1]$. The classes $\mathscr{B}_1^2 = \{B \times [0, 1], B \in \mathscr{B}\}$ and $\mathscr{B}_2^2 = \{[0, 1] \times B, B \in \mathscr{B}\}$ are sub-σ-algebras of $\mathscr{B}^2$ (Fig. 1.3). It follows that $\mathscr{B}_1^2$ and $\mathscr{B}_2^2$ are independent under the product Lebesgue measure λ^2.

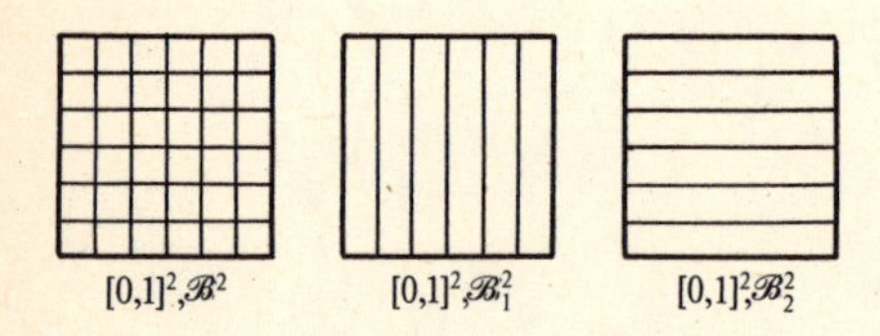

Fig. 1.3. The σ-algebras $\mathscr{B}^2$, $\mathscr{B}_1^2$ and $\mathscr{B}_2^2$

If $X : \Omega \to \mathbb{R}$ is any random variable, then

$$A_X = \{X^{-1} B, B \in \mathscr{B}\} \subset \mathscr{A}$$

is a sub-σ-field of $\mathscr{A}$. Now, if X_1 and X_2 are random variables inducing (sub-σ-fields of) $\mathscr{B}_1^2$ and $\mathscr{B}_2^2$ in $\mathscr{B}^2$, then X_1 and X_2 are independent.

Without elaboration, the following are direct consequences of the independence definition.

a) A set of r.v.'s $X^T = [X_1 \ldots X_n]$ is independent if and only if the joint probability distribution function of X satisfies

$$F_X(x) = F_{X_1}(x_1) F_{X_2}(x_2) \ldots F_{X_n}(x_n) \tag{1.45}$$

for all sets of real numbers $\{x_1, \ldots, x_n\}$. An equivalent statement is that the joint characteristic function ϕ_X of X satisfies

$$\phi_X(\boldsymbol{u}) = \phi_{X_1}(u_1)\,\phi_{X_2}(u_2)\ldots\phi_{X_n}(u_n) \tag{1.46}$$

for all $\boldsymbol{u}^T = [u_1, \ldots, u_n] \in \mathbb{R}^n$.

b) Subsets of independent sets of random variables are independent.

c) Let $f_s : \mathbb{R} \to \mathbb{R}$ be a Borel function for each $s \in S$. If a set of r.v.'s $\{X_s,\ s \in S\}$ is independent, then the set of r.v.'s $\{f_s \cdot X_s,\ s \in S\}$ is also independent.

d) If $C_1 \subset \mathscr{A}$ and $C_2 \subset \mathscr{A}$ are classes such that $\{C_1, C_2\}$ is independent, then if $A_1 \in C_1$ and $A_2 \in C_2$, the indicator functions i_{A_1} and i_{A_2} are independent random variables. Also, since $i_{A_1} \cdot i_{A_2} = i_{(A_1 \cap A_2)}$ and $\mathrm{E}\, i_A = PA$ for each $A \in \mathscr{A}$,

$$\mathrm{E}\{i_{A_1} \cdot i_{A_2}\} = \mathrm{E}\{i_{(A_1 \cap A_2)}\} = P(A_1 \cap A_2) = PA_1 PA_2 = \mathrm{E}\, i_{A_1} \mathrm{E}\, i_{A_2}\,.$$

e) If $X = \sum_{j=1}^{m} a_j\, i_{A_j}$ and $Y = \sum_{k=1}^{n} b_k\, i_{B_k}$, a_j and b_k in $\mathbb{R}$, $A_j \in C_1$, $B_k \in C_2$, and C_1 and C_2 as in (d), then, according to (d),

$$\begin{aligned}\mathrm{E}\{XY\} &= \mathrm{E}\left\{\sum_{j=1}^{m} a_j i_{A_j} \sum_{k=1}^{n} b_k i_{B_k}\right\}\\ &= \sum_{j=1}^{m}\sum_{k=1}^{n} a_j b_k \mathrm{E}\{i_{A_j \cap B_k}\} = \sum_{j=1}^{m}\sum_{k=1}^{n} a_j b_k PA_j PB_k\\ &= \left\{\sum_{j=1}^{m} a_j PA_j\right\}\left\{\sum_{k=1}^{n} b_k PB_k\right\} = \mathrm{E}X\,\mathrm{E}Y\,.\end{aligned}$$

The next point is a generalization of (e).

f) If a set of r.v.'s $\{X_1, \ldots, X_n\}$ is independent and if $\mathrm{E}X_i$, $i = 1, \ldots, n$, is finite, then $\mathrm{E}\{X_1 \ldots X_n\}$ exists and

$$\mathrm{E}\{X_1 \ldots X_n\} = \mathrm{E}X_1 \mathrm{E}X_2 \ldots \mathrm{E}X_n\,.$$

In particular, if r.v.'s X and Y are independent and have finite expectations, then

$$\mathrm{E}\{XY\} = \mathrm{E}X\,\mathrm{E}Y\,.$$

Moreover, in the case of at least one centered random variable, say $\mathrm{E}X = 0$, we have

$$\mathrm{E}\{XY\} = \mathrm{E}X\,\mathrm{E}Y = 0\,.$$

When $\mathrm{E}\{XY\} = 0$, X and Y are termed *orthogonal*.

1.3.2 The Gaussian N Vector and Gaussian Manifolds

Let $\{\Omega, \mathscr{A}, P\}$ be a probability space and consider r.v.'s $X_i : \Omega \to \mathbb{R}$, $i = 1, 2, \ldots, N$. The random vector $X^T = [X_1 \ldots X_N]$ is *Gaussian* or *normally* distributed if the joint probability density function of the random variables X_i, $i = 1, \ldots, N$, is given by

$$f_X(x) = \left(\frac{1}{2\pi}\right)^{N/2} \frac{1}{|C|^{1/2}} \exp\left[-\frac{1}{2}(x-m)^T C^{-1}(x-m)\right], \qquad (1.47)$$

where $x^T = [x_1 \ldots x_N] \in \mathbb{R}^N$, $m^T = [m_1 \ldots m_N] \in \mathbb{R}^N$, and $C = [c_{ij}]$ is a symmetric $N \times N$ real matrix with positive eigenvalues.

It can be easily verified that

$$\int_{-\infty}^{\infty} \ldots \int_{-\infty}^{\infty} f_X(x)\, dx = 1 \quad \text{and}$$

$$\mathrm{E}X = m \qquad (1.48)$$

$$\mathrm{E}\{(X - m)(X - m)^T\} = C\,. \qquad (1.49)$$

Thus, C is the *covariance matrix* of X and (1.48, 49) show that a Gaussian distribution is completely specified by the mean vector m and the covariance matrix C.

Following (1.43), the joint characteristic function of X can be shown to be

$$\phi_X(u) = \mathrm{E}\{e^{iu^TX}\} = \exp(iu^Tm - \tfrac{1}{2}u^TCu)\,, \qquad (1.50)$$

where $u^T = [u_1 \ldots u_N] \in \mathbb{R}^N$.

As a covariance matrix, C is positive semidefinite and hence may have zero eigenvalues. Thus, if we drop the condition that the eigenvalues of C be positive, C may be singular. If so, the components of X are linearly dependent and the joint probability density function given by (1.47) no longer exists. Hence, a preferred definition for normally distributed random variables is through its joint characteristic function, which always has meaning. We can state that the random vector X is normally distributed if its characteristic function ϕ_X is given by (1.50), where C is a symmetric $N \times N$ matrix with real entries and nonnegative eigenvalues.

We now cite some important properties associated with a normally distributed random vector. First, let X be normally distributed and consider

$$Y = AX\,, \qquad (1.51)$$

where $A = [a_{ij}]$ is a $K \times N$ real matrix. It is seen that

$$\phi_Y(\boldsymbol{v}) = \mathrm{E}\{e^{i\boldsymbol{v}^TY}\} = \mathrm{E}\{e^{i(A^T\boldsymbol{v})^TX}\} = \phi_X(A^T\boldsymbol{v})$$

$$= \exp(i\boldsymbol{v}^TA\boldsymbol{m} - \tfrac{1}{2}\boldsymbol{v}^TACA^T\boldsymbol{v}), \quad \boldsymbol{v} \in \mathbb{R}^K .$$

Hence, the random K vector Y is also normally distributed with mean $A\boldsymbol{m}$ and covariance matrix ACA^T. As a special case, the above result shows that each component X_j of X is normal with

$$\phi_{X_j}(u_j) = \exp(im_ju_j - \tfrac{1}{2}c_{jj}u_j^2), \quad j = 1, 2, \ldots, N,$$

in accordance with (1.35).

The results derived above can be restated as follows. The *linear hull* of the set $\{X_1, \ldots, X_N\}$ of components of a normally distributed N vector X in the linear space of random variables defined on $\{\Omega, \mathcal{A}, P\}$ is a *Gaussian manifold*; each finite ordered subset of this linear hull is a Gaussian vector. The manifold discussed here is finite-dimensional with dimension equal to or less than N.

Suppose that normally distributed r.v.'s X_i, $i = 1, 2, \ldots, N$, are centered and orthogonal, i.e.,

$$\mathrm{E}X_i = 0$$

$$\mathrm{E}\{X_iX_j\} = 0, \quad i \neq j.$$

Then the covariance matrix C takes the form

$$C = \begin{bmatrix} \mathrm{E}X_1^2 & & 0 \\ & \ddots & \\ 0 & & \mathrm{E}X_N^2 \end{bmatrix}$$

and (1.50) gives

$$\phi_X(\boldsymbol{u}) = \exp\left(-\tfrac{1}{2}\boldsymbol{u}^TC\boldsymbol{u}\right)$$

$$= \exp\left(-\tfrac{1}{2}EX_1^2u_1^2\right) \ldots \exp\left(-\tfrac{1}{2}EX_N^2u_N^2\right)$$

$$= \phi_{X_1}(u_1) \ldots \phi_{X_N}(u_N).$$

According to (1.46), the above result implies that r.v.'s $X_1, \ldots, X_N$ are independent. Hence, in the centered Gaussian case, orthogonality is equivalent to independence, a property not enjoyed by many other distributions.

1.4 Stochastic Processes

Given a system of r.v.'s X_i, $i = 1, \ldots, n$, it leads to the concept of a stochastic process by allowing the index i to range over an infinite set.

Let $\{\Omega, \mathscr{A}, P\}$ be a probability space, $\{\mathbb{R}, \mathscr{B}\}$ be the real line with Borel algebra $\mathscr{B}$, and I be an interval of $\mathbb{R}$ whose elements will be denoted by "t", standing for "time". A *stochastic process* (s.p.) or a *random process* is a set $\{X_t, t \in I\}$ such that, at each fixed $t \in I$, X_t is a random variable (i.e., it is a measurable mapping $X_t : \Omega \to \mathbb{R}$). The elements X_t will be called random variables of the process.

In what follows, we shall use the notation

$$X(t)\,, \qquad t \in I$$

to represent the stochastic process given above. It may be seen as a mapping of I into the space of all random variables defined on $\{\Omega, \mathscr{A}, P\}$. In this notation, the random variables of the process are the function values of $X(t)$, to be interpreted as measurable mappings $X(t) : \Omega \to \mathbb{R}$ for each fixed $t \in I$. Hence, at each fixed $t \in I$, $X(t)$ is a set of correspondences

$$\{(\omega, X(t)(\omega)); \omega \in \Omega\}$$

or simply

$$\{(\omega, X(t, \omega)), \omega \in \Omega\}\,.$$

With $X(t, \omega)$ defined for each $t \in I$ and $\omega \in \Omega$, it may be seen as a mapping $X : I \times \Omega \to \mathbb{R}$ with the property that its "sections" $X(t)$ at fixed $t \in I$ are random variables.

Given $X(t, \omega)$, $(t, \omega) \in I \times \Omega$, we may also single out a set of correspondences

$$\{(t, X(t, \omega)), t \in I\}$$

for fixed $\omega \in \Omega$. They are mappings of I into $\mathbb{R}$ and are just ordinary real functions. They are called *trajectories (sample functions, samples, realizations)* of the process.

Given a s.p. $X(t)$ defined in the sense outlined above, joint probability distributions of all finite sets of random variables associated with it can be derived with the following properties.

a) They are consistent in the sense that if S, S' and S'' are finite subsets of I such that $S \subset S' \cap S''$, the marginal probability distribution function of $\{X(t), t \in S\}$ derived from the probability distribution function of $\{X(t),$

$t \in S'\}$ is identical to that derived from the probability distribution function of $\{X(t), t \in S''\}$.

b) They are symmetric in the sense that if $i_1, i_2, \ldots, i_n$ is an arbitrary permutation of $1, 2, \ldots, n$, then

$$F_{X_{i_1} \ldots X_{i_n}}(x_{i_1}, \ldots, x_{i_n}) = F_{X_1 \ldots X_n}(x_1, \ldots, x_n) .$$

Before proceeding, let us note that the definition of a stochastic process as given above (i.e., as a function of two "variables" t and ω) is not completely natural. Since we are interested in making probability statements about a stochastic process, it would be more natural to define a stochastic process by means of a set of probability distribution functions or by other means leading directly to these distributions. A connection between a process represented by a function of t and ω and its probability distributions is contained in the following result from *Kolmogorov* and *Daniell* [1.1].

Given a complete set of finite-dimensional distribution functions with t ranging through $I \subset \mathbb{R}$, then, if these distributions satisfy the aforementioned conditions (a) and (b) of consistency and symmetry, there are (special) probability spaces $\{\Omega, \mathcal{A}, P\}$ and mappings

$$X : I \times \Omega \to \mathbb{R}$$

whose sections $X(t)$ at $t \in I$ are measurable and enjoy the prescribed finite-dimensional distribution functions.

As in the situation encountered in the random N-vector case, this result is not unique. But now the consequences are much more serious as we may show in the following example.

Example 1.5. Consider a s.p. $X(t)$, $t \in [0, T]$, with

$$P\{X(t) = 0\} = 1 \quad \text{for each fixed} \quad t \in [0, T] .$$

We shall construct several possible functions $X(t, \omega)$ with the prescribed distribution.

Case A. A simple model for this process is obtained by letting Ω consist of one point ω only. Then $\mathcal{A} = \{\emptyset, \Omega\}$, $P\Omega = 1$ and $P\emptyset = 0$. A suitable representation of the process $X(t)$ is the function

$$X(t, \omega) = 0 , \quad t \in [0, T] .$$

Case B. The probability space considered in Case A is not useful except in trivial situations. Let us now consider the space $\{[0, 1], \mathcal{B}, \lambda\}$ as a probability space (Sect. 1.1.1). This space is likewise not very useful but large enough to show some specific difficulties in the construction of $X(t, \omega)$.

Clearly, the random variables of

$$X' : [0, T] \times [0, 1] \rightarrow \mathbb{R}$$

defined by

$$X'(t, \omega) = 0 \quad \text{for all} \quad (t, \omega) \in [0, T] \times [0, 1]$$

have the prescribed distribution. The trajectories (sections of $X'(t, \omega)$ at fixed $\omega \in [0, 1]$) are constant functions of t on $[0, T]$ as shown in Fig. 1.4 and hence continuous with probability one.

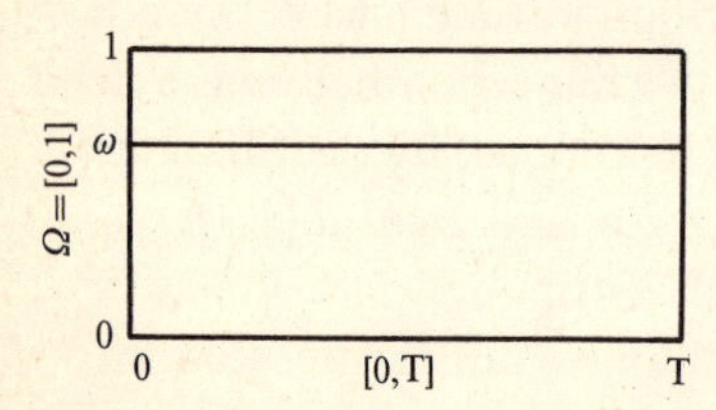

Fig. 1.4. The representation $X'(t, \omega)$

Case C. Consider the same probability space as in Case B, and the mapping

$$X'' : [0, T] \times [0, 1] \rightarrow \mathbb{R}$$

in which $X''(t, \omega) = 0$ everywhere except at the points (t, ω) on the lines drawn in Fig. 1.5, and $X''(t, \omega) = 1$ at these points. Now, the random variables of $X''(t, \omega)$ are a.s. identical to those of $X'(t, \omega)$ in Case B and have the prescribed distribution. But the trajectories of $X''(t, \omega)$ are discontinuous functions of t on $[0, T]$ with probability one.

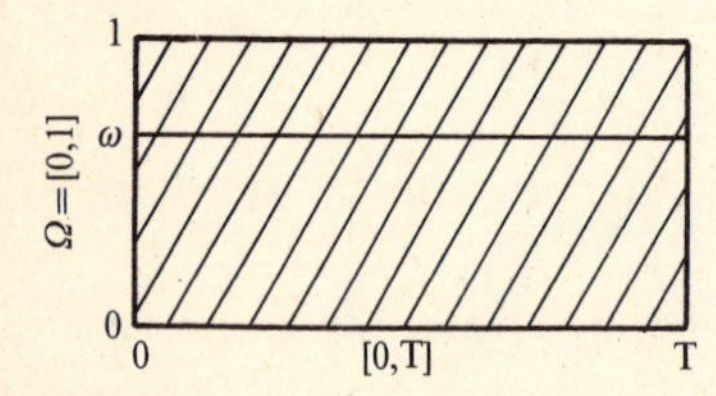

Fig. 1.5. The representation $X''(t, \omega)$

To circumvent difficulties such as nonuniqueness and other problems in trajectory specification and measurability, *Doob* [1.5] proposed that only so-called separable representations $X(t, \omega)$ of X are admissible. Roughly speaking, a separable representation is a function $X(t, \omega)$ whose random variables have the prescribed probability distribution *and* whose sections at ω are as smooth as possible. Under the restriction of separable representation, it turns out to be possible to define the trajectories of a process. While this point is important in the mathematical construction of stochastic processes, we shall not elaborate further as we are mainly concerned with processes having only continuous trajectories.

In closing, we write a vector s.p. $\boldsymbol{X}(t)$ in the form

$$\boldsymbol{X}(t) = \begin{bmatrix} X_1(t) \\ \vdots \\ X_N(t) \end{bmatrix}, \quad t \in I \tag{1.52}$$

with components as s.p. $X_i(t)$, $t \in I$, $i = 1, 2, \dots, N$. The vector s.p. $\boldsymbol{X}(t)$ is defined if the set of probability distribution functions of all finite sets of r.v.'s $\{X_1(t_1), \dots, X_N(t_1), \dots, X_1(t_M), \dots X_N(t_M)\}$, $\{t_1, \dots, t_M\} \subset I$, is given and satisfies conditions (a) and (b) of consistency and symmetry stated earlier.

1.4.1 The Hilbert Space $L_2(\Omega)$

We are concerned with an important class of stochastic processes called *second-order* processes. The Hilbert space $L_2(\Omega)$ occupies a central place in the discussion of properties associated with second-order stochastic processes. Hence, some preliminary remarks concerning the space $L_2(\Omega)$ are in order.

Let $\{\Omega, \mathscr{A}, P\}$ be a probability space and let $X : \Omega \to \mathbb{R}$ be a *second-order* random variable, i.e., $\mathrm{E}X^2 < \infty$. Let us first consider some properties of the set S of all second-order random variables defined on Ω, namely,

$$S = \{X : \Omega \to \mathbb{R}; \quad X \text{ is measurable and } \quad \mathrm{E}X^2 < \infty\} .$$

Linear Vector Space. It is easy to show that if X and Y are in S and $c \in \mathbb{R}$, we have

$$1) \quad cX \in S \tag{1.53}$$

since cX is measurable and $\mathrm{E}\{cX\}^2 = c^2\mathrm{E}X^2 < \infty$, and

$$2) \quad X + Y \in S \tag{1.54}$$

since $X + Y$ is measurable and $\mathrm{E}\{X + Y\}^2 < \infty$. This inequality is seen as follows:

$$(X + Y)^2 \leq 2X^2 + 2Y^2 \Rightarrow \mathrm{E}\{X + Y\}^2 \leq 2\mathrm{E}X^2 + 2\mathrm{E}Y^2 < \infty .$$

Since S is a subset of the space of all random variables defined on Ω, (1.53, 54) imply that S is a *linear vector space*.

Inner Product. If X and Y are elements of S, XY is measurable and $|XY| \leqslant \frac{1}{2}X^2 + \frac{1}{2}Y^2$, implying $\mathrm{E}\{|XY|\} < \infty$ and hence $|\mathrm{E}\{XY\}| < \infty$. Let us use the notation

$$\mathrm{E}\{XY\} = \langle X, Y \rangle . \tag{1.55}$$

It is an *inner product* in S as it satisfies the following inner product properties

$$\left.\begin{array}{l} \langle X, Y \rangle = \langle Y, X \rangle \\ \langle (X + Y), Z \rangle = \langle X, Z \rangle + \langle Y, Z \rangle \\ \langle cX, Y \rangle = c\langle X, Y \rangle \\ \langle X, X \rangle \geqslant 0 ; \quad \langle X, X \rangle = 0 \Leftrightarrow X = 0 \end{array}\right\} \tag{1.56}$$

for any X, Y and Z in S and any $c \in \mathbb{R}$. The first three conditions are easily shown to be satisfied. The fourth one, however, is not true in a strict sense since

$$\mathrm{E}X^2 = 0 \Leftrightarrow X(\omega) = 0 \text{ a.s.}$$

However, partitioning S into equivalence classes of a.s. identical random variables (Sect. 1.2) and defining the operations on the classes by means of the operations on the random variables, all properties given by (1.56) hold for the equivalence classes of second-order random variables. With the equivalence classes in mind, $\mathrm{E}\{XY\}$ is an inner product. If $\mathrm{E}\{XY\} = 0$, X and Y are called orthogonal and we write $X \perp Y$.

If $X_1, X_2, \ldots, X_n$ in S are mutually orthogonal, we obtain the Pythagorean law

$$\langle (X_1 + \ldots + X_n), (X_1 + \ldots + X_n) \rangle = \langle X_1, X_1 \rangle + \ldots + \langle X_n, X_n \rangle . \tag{1.57}$$

Cauchy, Bunyakowsky, and Schwarz Inequality. If X and Y are elements of S, we have, for all $x \in \mathbb{R}$,

$$\mathrm{E}\{xX + Y\}^2 \geqslant 0 \quad \text{i.e.,}$$

$$(\mathrm{E}X^2)x^2 + 2(\mathrm{E}\{XY\})x + \mathrm{E}Y^2 \geqslant 0 .$$

Hence, the discriminant of this quadratic form in x is not positive and we obtain Cauchy's inequality

$$(\mathrm{E}\{XY\})^2 \leqslant \mathrm{E}X^2\mathrm{E}Y^2 . \tag{1.58}$$

Norm. Let

$$\langle X, X \rangle^{1/2} = \|X\| . \tag{1.59}$$

It is a *norm* as the following conditions are satisfied:

$$\left.\begin{aligned}&\|X+Y\| \leqslant \|X\|+\|Y\| \\ &\|cX\| = |c|\,\|X\| \\ &\|X\| \geqslant 0\,; \quad \|X\| = 0 \Leftrightarrow X = 0\,.\end{aligned}\right\} \tag{1.60}$$

The first property is true if X and Y are in S, since by (1.58)

$$\begin{aligned}\|X+Y\|^2 &= \mathrm{E}X^2 + 2\mathrm{E}\{XY\} + \mathrm{E}Y^2 \leqslant \|X\|^2 + 2\|X\|\,\|Y\| + \|Y\|^2 \\ &= \{\|X\| + \|Y\|\}^2\,.\end{aligned}$$

It is easily seen that the second property also holds and it follows from discussion on the inner product that for a r.v. X in S, $\|X\|$ has all the norm properties when the equivalence classes are taken into account.

Using the norm notation, the Pythagorean law of (1.57) takes the form

$$\|X_1 + \ldots + X_n\|^2 = \|X_1\|^2 + \ldots + \|X_n\|^2 \tag{1.61}$$

when $X_1, X_2, \ldots, X_n$ are mutually orthogonal.

The Cauchy inequality reads

$$|\mathrm{E}\{XY\}| \leqslant \|X\|\,\|Y\|\,. \tag{1.62}$$

Distance. Finally, define the *distance* between X and Y by

$$d(X, Y) = \|X - Y\|\,. \tag{1.63}$$

We can show that for any X and Y in S and taking into account equivalence classes in S, $d(X, Y)$ possesses all the usual distance properties, i.e.,

$$\left.\begin{aligned}&d(X, Y) = d(Y, X) \\ &d(X, Y) \leqslant d(X, Z) + d(Z, Y) \\ &d(X, Y) \geqslant 0\,; \quad d(X, Y) = 0 \Leftrightarrow X = Y\,.\end{aligned}\right\} \tag{1.64}$$

Convergence. A sequence

$$\{X_n\}_{n \in \mathbb{N}}\,, \qquad X_n \in S$$

is said to be convergent with limit $X \in S$ if $d(X_n, X) \to 0$ as $n \to \infty$, i.e., if

$$\|X_n - X\| \to 0$$

as $n \to \infty$. It is called a Cauchy, or fundamental, sequence if

$$\|X_m - X_n\| \to 0$$

as m and n tend to infinity independently of each other.

Clearly each convergent sequence in S is a Cauchy sequence. Conversely, it can be shown that each Cauchy sequence in S converges, i.e., has a limit in S; or, in other words, S is a complete space.

As a complete inner product space, S is a *Hilbert space*. In the sequel, the space S of all second-order random variables will be denoted by $L_2(\Omega, \mathscr{A}, P)$, $L_2(\mathscr{A})$, or $L_2(\Omega)$ when the meaning is clear.

In closing, we remark that the indicator function i_A of $A \in \mathscr{A}$ belongs to $L_2(\Omega)$. This is true since

$$\mathrm{E}\, i_A^2 = \mathrm{E}\, i_A = PA < \infty\,. \tag{1.65}$$

We also see that if $X \in L_2(\Omega)$, $|\mathrm{E}X| < \infty$ since by (1.62)

$$|\mathrm{E}X| = |\mathrm{E}\{i_\Omega X\}| \leqslant \|i_\Omega\|\,\|X\| = \|X\|\,. \tag{1.66}$$

Moreover, $X - \mathrm{E}X \in L_2(\Omega)$ since it may be seen as a linear combination of the elements X and $(\mathrm{E}X)\, i_\Omega$ in S. We thus have

$$\left.\begin{array}{l} \mathrm{E}X^2 < \infty \Rightarrow |\mathrm{E}X| < \infty \\ |\mathrm{E}X| \leqslant \|X\| \\ \mathrm{E}X^2 < \infty \Rightarrow X - \mathrm{E}X \in L_2(\Omega) \quad \text{and} \quad \mathrm{E}\{X - \mathrm{E}X\} = 0 \\ \mathrm{E}\{X - \mathrm{E}X\}^2 = \mathrm{E}X^2 - (\mathrm{E}X)^2\,. \end{array}\right\} \tag{1.67}$$

If $A = \{\omega \in \Omega \,\big|\, |X(\omega)| \geqslant \varepsilon\}, \quad \varepsilon \in \mathbb{R}^+,$ then

$$|X| \geqslant \varepsilon\, i_A$$

and hence

$$\mathrm{E}X^2 \geqslant \varepsilon^2 P(A)\,, \quad \text{i.e.,}$$

$$P\{|X| \geqslant \varepsilon\} \leqslant \frac{l}{\varepsilon^2}\, \mathrm{E}X^2\,. \tag{1.68}$$

This inequality is called the *Chebyshev inequality*.

1.4.2 Second-Order Processes

Let $I \subset \mathbb{R}$ and let $\{\Omega, \mathscr{A}, P\}$ be a probability space. A process $\{X(t) : \Omega \to \mathbb{R},\, t \in I\}$ having the property that $X(t) \in L_2(\Omega)$ for each $t \in I$ is called a *second-order* process. Since it is a one-parameter set of second-order random variables, it may be seen as a "curve" in the Hilbert space $L_2(\Omega)$. Results in

Sect. 1.4.1 show that $\mathrm{E}X(t)$, $t \in I$, is a finite real function and $X(t) - \mathrm{E}X(t)$, $t \in I$, is a centered second-order process.

The *correlation function* of $X(t)$, $t \in I$, is defined by

$$\mathrm{E}\{X(s)\,X(t)\}\,, \quad (s,t) \in I^2\,.$$

As seen from Cauchy's inequality

$$|\mathrm{E}\{X(s)\,X(t)\}| \leqslant \|X(s)\|\,\|X(t)\|\,,$$

the correlation function of a second-order process is a finite mapping of I^2 into $\mathbb{R}$.

The expectation

$$\mathrm{E}\{[X(s) - \mathrm{E}X(s)][X(t) - \mathrm{E}X(t)]\} = \mathrm{E}\{X(s)X(t)\} - \mathrm{E}X(s)\,\mathrm{E}X(t)\,, \quad (s,t) \in I^2 \tag{1.69}$$

is called the *covariance function* of $X(t)$.

Given two second-order s.p.'s $X(t)$ and $Y(t)$, $t \in I$, the expectation

$$\mathrm{E}\{X(s)\,Y(t)\}\,, \quad (s,t) \in I^2 \tag{1.70}$$

is similarly a finite mapping of I^2 into $\mathbb{R}$ and is called the *cross-correlation function* of $X(s)$ and $Y(t)$. Their *cross-covariance function* is

$$\mathrm{E}\{[X(s) - \mathrm{E}X(s)][Y(t) - \mathrm{E}Y(t)]\} = \mathrm{E}\{X(s)\,Y(t)\} - \mathrm{E}X(s)\,\mathrm{E}Y(t)\,, \quad (s,t) \in I^2\,. \tag{1.71}$$

The definitions given above can be extended to the N – r.v. case in a straightforward fashion. Consider a second-order vector s.p.

$$\boldsymbol{X}(t) = \begin{bmatrix} X_1(t) \\ \vdots \\ X_N(t) \end{bmatrix}, \quad t \in I \tag{1.72}$$

defined as one whose component processes $X_i(t)$, $i = 1, 2, \ldots, N$, are second order. Its *correlation (function) matrix* is given by

$$\mathrm{E}\{\boldsymbol{X}(s)\boldsymbol{X}^T(t)\} = \begin{bmatrix} \mathrm{E}\{X_1(s)X_1(t)\} \ldots \mathrm{E}\{X_1(s)X_N(t)\} \\ \vdots \qquad\qquad \vdots \\ \mathrm{E}\{X_N(s)X_1(t)\} \ldots \mathrm{E}\{X_N(s)X_N(t)\} \end{bmatrix}, \quad (s,t) \in I^2\,. \tag{1.73}$$

The diagonal terms give the correlation functions of the component processes and the nondiagonal terms are their cross-correlation functions.

Similarly, the *covariance (function) matrix* of $\boldsymbol{X}(t)$ is

$$\mathrm{E}\{[\boldsymbol{X}(s)-\mathrm{E}\boldsymbol{X}(s)][\boldsymbol{X}(t)-\mathrm{E}\boldsymbol{X}(t)]^T\} \,. \tag{1.74}$$

Analogously, if the components $Y_1(t), \ldots, Y_M(t)$ of

$$\boldsymbol{Y}(t) = \begin{bmatrix} Y_1(t) \\ \vdots \\ Y_M(t) \end{bmatrix}, \quad t \in I$$

are also second order, the $N \times M$ matrices

$$\mathrm{E}\{\boldsymbol{X}(s)\boldsymbol{Y}^T(t)\}\,, \quad (s,t) \in I^2 \tag{1.75}$$

and

$$\mathrm{E}\{[\boldsymbol{X}(s)-\mathrm{E}\boldsymbol{X}(s)][\boldsymbol{Y}(t)-\mathrm{E}\boldsymbol{Y}(t)]^T\}\,, \quad (s,t) \in I^2 \tag{1.76}$$

are called, respectively, their *(cross-) correlation (function) matrix* and *(cross-) covariance (function) matrix*.

1.4.3 The Gaussian Process

The discussion in Sect. 1.3.2. concerning Gaussian vectors and manifolds leads directly to a class of important stochastic processes called *Gaussian processes*.

Let $\{\Omega, \mathcal{A}, P\}$ be a probability space, $I \subset \mathbb{R}$, and $X : I \to L_2(\Omega)$ be a second-order process. If, for each finite subset $S = \{t_1, \ldots, t_N\}$ of I, the characteristic function of the vector

$$\boldsymbol{X}(S) = \begin{bmatrix} X(t_1) \\ \vdots \\ X(t_N) \end{bmatrix}$$

is in the form

$$\phi_X(u) = \exp[\mathrm{i}\boldsymbol{u}^T\boldsymbol{m}(S) - \tfrac{1}{2}\boldsymbol{u}^T C(S)\boldsymbol{u}]\,, \tag{1.77}$$

the process X is called a *Gaussian* stochastic process. In (1.77), $\boldsymbol{m}(S) = \mathrm{E}\boldsymbol{X}(S)$, $C(S) = \mathrm{E}\{[\boldsymbol{X}(S)-\boldsymbol{m}(S)][\boldsymbol{X}(S)-\boldsymbol{m}(S)]^T\}$ and $\boldsymbol{u}^T = [u_1 \ldots u_N] \in \mathbb{R}^N$.

The set of probability distributions corresponding to the characteristic functions given above satisfies the consistency and symmetry conditions in Sect. 1.4.

Let us consider the linear hull $L\{X(t),\ t \in I\}$, i.e., the set of all linear combinations of finitely many random variables $X(t)$ of the process X. It is a Gaussian manifold but is not of finite dimension in general and is in general not a *closed* subspace of $L_2(\Omega)$. It can be shown that the closure of $L\{X(t),\ t \in I\}$ in $L_2(\Omega)$ is also Gaussian.

A second-order vector process

$$\boldsymbol{X}(t) = \begin{bmatrix} X_1(t) \\ \vdots \\ X_N(t) \end{bmatrix}, \quad t \in I$$

is Gaussian if, for each finite set $S = \{t_i, \ldots, t_m\} \subset I$, the characteristic function of the vector

$$\boldsymbol{X}(S) = \begin{bmatrix} X_1(t_1) \\ \vdots \\ X_N(t_1) \\ \vdots \\ X_1(t_m) \\ \vdots \\ X_N(t_m) \end{bmatrix}$$

is given by

$$\phi_X(\boldsymbol{u}) = \exp\left[\mathrm{i}\boldsymbol{u}^T\boldsymbol{m}(S) - \tfrac{1}{2}\boldsymbol{u}^T C(S)\,\boldsymbol{u}\right], \tag{1.78}$$

where $\boldsymbol{m}(S) = \mathrm{E}\boldsymbol{X}(S)$, $C(S) = \mathrm{E}\{[\boldsymbol{X}(S) - \boldsymbol{m}(S)][\boldsymbol{X}(S) - \boldsymbol{m}(S)]^T\}$, and $\boldsymbol{u}^T = [u_1 \ldots u_{Nm}] \in \mathbb{R}^{Nm}$.

Hence, the distribution of a Gaussian vector process $\boldsymbol{X}(t)$, $t \in I$, is specified by its mean $\mathrm{E}\boldsymbol{X}(t)$, and its covariance function matrix $\mathrm{E}\{[\boldsymbol{X}(s) - \boldsymbol{m}(s)][\boldsymbol{X}(t) - \boldsymbol{m}(t)]^T\}$, $\quad (s, t) \in I^2$.

Again, the set of corresponding probability distributions satiesfies the consistency and symmetry conditions stated in Sect. 1.4.

The linear hull of $\{X_i(t), i = 1, \ldots, N; t \in I\}$ and its closure in $L_2(\Omega)$ are Gaussian spaces.

1.4.4 Brownian Motion, the Wiener-Lévy Process and White Noise

In 1828, Robert Brown, a botanist, observed that small particles immersed in a liquid move irregularly. This phenomenon, correctly described by Brown as a result of impact of molecules in the liquid, is called *Brownian motion*. A mathematical model of Brownian motion is the Wiener-Lévy process. While this process has undesirable properties such as sample nondifferentiability, it remains very important due to its own interesting as well as practical features. Moreover, white noise, a process in extensive technological usage, can be represented as a "formal" derivative of the Wiener-Lévy process.

In what follows, a characterization of the so-called standard Wiener-Lévy process is given. More detailed discussion can be found in [1.5], for example.

Let $\{\Omega, \mathscr{A}, P\}$ be a suitable probability space, let $I = [0, T] \subset \mathbb{R}$ and let

$$W = \{W(t)\ ,\ t \in I\}$$

be a stochastic process whose random variables are mappings $W(t) : \Omega \to \mathbb{R}$. The process W is the *standard Wiener-Lévy process* if the following six conditions are satisfied.

a) $W(0) = 0$ a.s.

b) The process W has independent increments on I, that is, for each set of mutually disjoint intervals of I, $\{[t_i, t_{i+1}),\ i = 1, 2, \ldots, n\}$, the *increments* of W on $[t_i, t_{i+1})$, $i = 1, 2, \ldots n$, defined as

$$\{W(t_{i+1}) - W(t_i)\}$$

are independent.

c) The process W is sample continuous, i.e., the trajectories of W are continuous mappings of I into $\mathbb{R}$ with probability one. In other words, at each fixed $\omega \in \Omega$ outside a set of probability zero, the function $W(t, \omega)$ is continuous in $t \in I$.

d) $\mathrm{E}W(t) = 0\ , \quad t \in I.$

e) The increments $\{W(s) - W(t)\}$ of W are stationary in the sense that the expectation

$$\mathrm{E}\{W(s) - W(t)\}^2 = \|W(s) - W(t)\|^2$$

is a function of $(s - t)$ only. This property together with the foregoing conditions leads to the result that

$$\mathrm{E}\{W(s) - W(t)\}^2 = c|s - t|\ , \tag{1.79}$$

where $c \geqslant 0$ is an arbitrary constant.

f) $c = 1$ in (1.79).

A number of interesting properties of the standard Wiener-Lévy process can be deduced from these conditions. A direct consequence of conditions a, b, and c is that W is a Gaussian process, and hence of second order. It can also be shown that W is continuous in mean square on I, meaning that

$$\|W(s) - W(t)\| \to 0 \quad \text{as} \quad s \to t\,, \qquad s, t \in I\,.$$

This type of continuity is discussed in Chap. 2.

With the addition of condition d, we also see that the increments $\{W(t_{i+1}) - W(t_i),\ i = 1, 2, \ldots n\}$ are orthogonal in $L_2(\Omega)$ since, if $i \neq j$,

$$\begin{aligned} &\mathrm{E}\{[W(t_{i+1}) - W(t_i)][W(t_{j+1}) - W(t_j)]\} \\ &\quad = \mathrm{E}\{W(t_{i+1}) - W(t_i)\}\,\mathrm{E}\{W(t_{j+1}) - W(t_j)\} = (0)\,(0) = 0\,. \end{aligned}$$

While W is sample continuous by definition, it can be shown that, with probability one, its trajectories are not of bounded variation on any interval of $[0, T]$ nor are they differentiable.

The covariance function of W is given by

$$\mathrm{E}\{W(s)\,W(t)\} = \min(s, t)\,, \qquad (s, t) \in [0, T]^2\,. \tag{1.80}$$

This can be seen by writing, with $0 \leqslant s \leqslant t \leqslant T$,

$$\begin{aligned} \mathrm{E}\{W(s)\,W(t)\} &= \mathrm{E}\{W(s)[W(s) + W(t) - W(s)]\} \\ &= \mathrm{E}W^2(s) + \mathrm{E}\{[W(s) - W(0)][W(t) - W(s)]\} = s + 0 = s\,. \end{aligned}$$

The vector process

$$\mathbf{W}^o(t) = \begin{bmatrix} W_1^o(t) \\ \vdots \\ W_N^o(t) \end{bmatrix}, \qquad t \in [0, T] \tag{1.81}$$

is called the *standard Wiener-Lévy N vector* if

$$\{W_i^o : [0, T] \to L_2(\Omega)\,; \qquad i = 1, \ldots, N\}$$

is an independent set of standard Wiener-Lévy processes. Due to independence of the increments of W_i^o and independence of W_i^o and W_j^o $(i \neq j)$, it is easily shown that

$$\mathrm{E}\mathbf{W}^o(t) = \begin{bmatrix} 0 \\ \vdots \\ 0 \end{bmatrix}, \qquad t \in [0, T] \tag{1.82}$$

and, if $0 \leqslant s \leqslant t \leqslant T$,

$$\mathrm{E}\{\mathbf{W}^o(s)\,\mathbf{W}^{oT}(t)\} = sI\,, \tag{1.83}$$

where I is the $N \times N$ identity matrix.

2. Calculus in Mean Square

Having introduced the concept of a stochastic process in Sect. 1.4, some elements of calculus in mean square, or m.s. calculus, are discussed in this chapter to the extent required for Chap. 3. Since the class of stochastic processes to be considered contains real functions as a special case, some of this development is also applicable to ordinary real functions.

As mentioned in Chap. 1, all random variables encountered in this development are understood to be defined on one and the same suitable probability space $\{\Omega, \mathscr{A}, P\}$. Furthermore, they are elements of the space $L_2(\Omega)$ as defined in Sect. 1.4.1. If two r.v.'s X and Y are a.s. equal, we usually write $X = Y$, omitting "a.s.".

2.1 Convergence in Mean Square

As in the development of ordinary calculus, we begin by introducing the concept of convergence in mean square (m.s. convergence) for a sequence of random variables. Some of the basic notions given below have been introduced in Sect. 1.4.1, and the proofs of some of the theorems are omitted but can be found in [2.1, 2], for example.

Let $\{X_n\}_{n\in N}$ and $\{Y_n\}_{n\in N}$ be sequences of elements of $L_2(\Omega)$ and let X and Y be elements of $L_2(\Omega)$.

Definition. A sequence $\{X_n\}_{n\in N}$ is said to be a *Cauchy sequence* if

$$\|X_m - X_n\| \to 0 \quad \text{as} \quad n, m \to \infty .$$

A sequence $\{X_n\}_{n\in N}$ is said to *converge in m.s.* to X if

$$\|X_n - X\| \to 0 \quad \text{as} \quad n \to \infty . \tag{2.1}$$

We recall that

$$\|X_n - X\| \to 0 \Leftrightarrow \mathrm{E}\{X_n - X\}^2 \to 0 .$$

Notationally, we can also write

$$X_n \xrightarrow{\text{m.s.}} X \quad \text{or} \quad \underset{n\to\infty}{\text{l.i.m.}} X_n = X\,, \tag{2.2}$$

where the symbol l.i.m. stands for "limit in mean (square)".

Theorem 2.1 (proof omitted). (a) Space $L_2(\Omega)$ is a complete space. That is, if $\{X_n\}_{n\in N}$ is a Cauchy sequence in $L_2(\Omega)$, then there is an element $X \in L_2(\Omega)$ such that $X_n \xrightarrow{\text{m.s.}} X$ as $n \to \infty$, see [2.1].

b) The limit X as given above is unique in the sense that if also $X_n \xrightarrow{\text{m.s.}} Y$, then $X = Y$ a.s.

c) A sequence in $L_2(\Omega)$ is m.s. convergent if and only if it is a Cauchy sequence.

In connection with m.s. convergence, we recall Cauchy's inequality (1.62)

$$|\mathrm{E}\{XY\}| \leqslant \|X\|\,\|Y\|$$

and note the following useful properties.

A) If $X_n \xrightarrow{\text{m.s.}} X$, then $\|X_n\| \to \|X\|$ and $\mathrm{E}X_n \to \mathrm{E}X$. This can be easily verified by noting that

$$|\,\|X_n\| - \|X\|\,| \leqslant \|X_n - X\| \to 0$$

and, as seen from (1.67),

$$|\mathrm{E}X_n - \mathrm{E}X| = |\mathrm{E}\{X_n - X\}| \leqslant \|X_n - X\| \to 0\,. \qquad \square$$

B) Continuity of Inner Product. If $X_n \xrightarrow{\text{m.s.}} X$ and $Y_m \xrightarrow{\text{m.s.}} Y$ as $n, m \to \infty$, then $\mathrm{E}\{X_n Y_m\} \to \mathrm{E}\{XY\}$. This result follows from Cauchy's inequality since

$$\begin{aligned}|\mathrm{E}\{XY\} - \mathrm{E}\{X_n Y_m\}| &\leqslant |\mathrm{E}\{(X - X_n)Y\}| + |\mathrm{E}\{X_n(Y - Y_m)\}| \\ &\leqslant \|X - X_n\|\,\|Y\| + \|X_n\|\,\|Y - Y_m\|\,,\end{aligned}$$

which approaches zero as $\|X_n\| \to \|X\| < \infty$ by virtue of (A). $\square$

C) Criterion for m.s. convergence. The sequence $\{X_n\}_{n\in N}$ is m.s. convergent if and only if $\mathrm{E}\{X_n X_m\}$ converges as $n \to \infty$ and $m \to \infty$ independently.

Proof. The "only if" part is a consequence of (B). For the "if" part, suppose $\mathrm{E}\{X_n X_m\} \to c$ as $n, m \to \infty$. Then,

$$\|X_n - X_m\|^2 = \mathrm{E}\{X_n - X_m\}^2 = \mathrm{E}X_n^2 - 2\,\mathrm{E}\{X_n X_m\} + \mathrm{E}X_m^2 \to c - 2c + c = 0\,.$$

Hence, $\{X_n\}_{n\in N}$ is a Cauchy sequence and it follows from Theorem 2.1 that it is m.s. convergent. □

Let I be an interval of the real line, and $X : I \to L_2(\Omega)$ be a second-order process whose random variables are denoted by

$$X(t)\ , \quad t \in I$$

and let $s \in I$, and $X_o \in L_2(\Omega)$.

Definition. $X(s) \xrightarrow{\text{m.s.}} X_o$ as $s \to s_o$ if

$$\|X(s) - X_o\| \to 0 \ .$$

If $s \to s_o$, it is tacitly assumed that $s \in I$ and $s \neq s_o$.

Theorem 2.2. $X(s)$ converges to some element of $L_2(\Omega)$ as $s \to s_o$ if and only if, for each sequence $\{s_n\}_{n\in N}$ in I converging to s_o, $\{X(s_n)\}_{n\in N}$ is a Cauchy sequence in $L_2(\Omega)$.

Proof. Consider the "only if" part and suppose $X(s) \xrightarrow{\text{m.s.}} X_o \in L_2(\Omega)$ as $s \to s_o$. Let $s_n \to s_o$ as $n \to \infty$. Then,

$$\forall \varepsilon > 0 \ \exists \ \delta > 0 : 0 < |s - s_o| < \delta \Rightarrow \|X(s) - X_o\| < \varepsilon \ .$$

Now, associated with δ there is a number N such that $n > N$ implies $0 < |s_n - s_o| < \delta$, and hence

$$\|X(s_n) - X_o\| < \varepsilon \quad \text{as} \quad n > N \ .$$

To show that the "if" part is true, we first see that for all sequences converging to s_o, the corresponding Cauchy sequences have the same limit. Suppose that $\{s_n\}_{n\in N}$ and $\{t_n\}_{n\in N}$ satisfy the above conditions and the sequences $\{X(s_n)\}_{n\in N}$ and $\{X(t_n)\}_{n\in N}$ converge in $L_2(\Omega)$ to, say, X_o and Y, respectively. We can show $X_o = Y$ by means of the following argument. Since the sequence $s_1, t_1, s_2, t_2, \ldots, s_n, t_n, \ldots$ also converges to s_o, $X(s_1)$, $X(t_1)$, $X(s_2)$, $X(t_2), \ldots, X(s_n)$, $X(t_n), \ldots$ is also a Cauchy sequence in $L_2(\Omega)$. Hence, for each $\varepsilon > 0$,

$$\|X_o - Y\| \leqslant \|X_o - X(s_n)\| + \|X(s_n) - X(t_n)\| + \|X(t_n) - Y\| < 3\,\varepsilon$$

if n is sufficiently large. We thus have $X_o = Y$.

Finally, we shall show $X(s) \xrightarrow{\text{m.s.}} X_o$ as $s \to s_o$, where X_o is the element defined above. Suppose $s \to s_o$ and $X(s)$ does not converge in m.s. to X_o.

This implies

$$\exists \varepsilon > 0 \; \forall \delta > 0 \; \exists s, \text{ such that } 0 < |s - s_o| < \delta \quad \text{and} \quad \|X(s) - X_o\| \geqslant \varepsilon \,.$$

Let $\{\delta_n > 0\}_{n \in N}$ be a sequence tending to 0 as $n \to \infty$. If we substitute the respective values δ_n for δ given above and denote the corresponding values of s by s_n, then $s_n \to s_o$ as $n \to \infty$, whereas $\|X(s_n) - X_o\| \geqslant \varepsilon$ entailing $X(s_n) \not\to X_o$ in m.s. as $n \to \infty$, contrary to the result stated above. Hence,

$$X(s) \xrightarrow{\text{m.s.}} X_o \quad \text{as} \quad s \to s_o \,. \qquad \square$$

Theorem 2.3. If X and Y are mappings of I into $L_2(\Omega)$ and if $X(s) \xrightarrow{\text{m.s.}} X_o$ and $Y(s) \xrightarrow{\text{m.s.}} Y_o$ as $s \to t$, then

$$aX(s) + bY(s) \xrightarrow{\text{m.s.}} aX_o + bY_o \quad \text{as} \quad s \to t \,, \tag{2.3}$$

where a and b are real numbers.

Proof. As $s \to t$

$$\|aX(s) + bY(s) - aX_o - bY_o\| \leqslant |a| \, \|X(s) - X_o\| + |b| \, \|Y(s) - Y_o\| \to 0 \,.$$

Hence, (2.3) follows. □

Theorem 2.4. Suppose $f : I \to \mathbb{R}$, $X : I \to L_2(\Omega)$. If $f(s) \to f_o$ and if $X(s) \xrightarrow{\text{m.s.}} X_o$ as $s \to t$, then

$$f(s)X(s) \xrightarrow{\text{m.s.}} f_o X_o \quad \text{as} \quad s \to t \,. \tag{2.4}$$

Proof. This result follows directly by noting that

$$\|f(s)X(s) - f_o X_o\| \leqslant |f(s)| \, \|X(s) - X_o\| + |f(s) - f_o| \, \|X_o\| \to 0$$

as $s \to t$. □

2.2 Continuity in Mean Square

Let $I \subset \mathbb{R}$ be an interval and let $X : I \to L_2(\Omega)$ be a second-order stochastic process whose random variables are denoted by $X(t)$, $t \in I$.

Definition. (a) X is *continuous in mean square* (m.s. *continuous) at* $t \in I$ if

$$\|X(s)-X(t)\| \to 0 \quad \text{as} \quad s \to t \quad \text{in} \quad I \, ; \tag{2.5}$$

b) X is m.s. *continuous on* I if it is m.s. continuous at each $t \in I$;

c) X is *uniformly* m.s. *continuous on* I if, for each $\varepsilon > 0$, there is a $\delta > 0$ such that

$$\|X(t_1) - X(t_2)\| < \varepsilon$$

for any t_1 and t_2 in I with $|t_1 - t_2| < \delta$.

Theorem 2.5. If $X : I \to L_2(\Omega)$ is m.s. continuous on I, where $I \subset \mathbb{R}$ is closed and bounded, then X is uniformly m.s. continuous on I.

Proof. Given $\varepsilon > 0$, then

$$\forall t \in I \; \exists \; \delta_t > 0 : s \in (t - \delta_t, t + \delta_t) \cap I \Rightarrow \|X(s) - X(t)\| < \frac{\varepsilon}{2} \, .$$

Now, $\{(t - \frac{1}{2}\delta_t, t + \frac{1}{2}\delta_t), t \in I\}$ is an open covering of the compact set I. Hence, it contains a finite covering of I, say,

$$\{(t_1 - \tfrac{1}{2}\delta_{t_1}, t_1 + \tfrac{1}{2}\delta_{t_1}), \ldots, (t_n - \tfrac{1}{2}\delta_{t_n}, t_n + \tfrac{1}{2}\delta_{t_n})\} \, . \tag{2.6}$$

Define $\delta_o = \min(\frac{1}{2}\delta_{t_1}, \ldots, \frac{1}{2}\delta_{t_n})$ and suppose s and t in I are such that $|s - t| < \delta_o$. Since $t \in I$, there is an element of the covering (2.6) containing t, say, $t \in (t_k - \frac{1}{2}\delta_{t_k}, t_k + \frac{1}{2}\delta_{t_k})$ as indicated in Fig. 2.1. Hence,

$$|s - t_k| \leqslant |s - t| + |t - t_k| \leqslant \delta_o + \tfrac{1}{2}\delta_{t_k} \leqslant \delta_{t_k}$$

and, therefore,

$$\|X(s) - X(t)\| \leqslant \|X(s) - X(t_k)\| + \|X(t_k) - X(t)\| < \frac{\varepsilon}{2} + \frac{\varepsilon}{2} = \varepsilon \, ,$$

and the proof is complete.

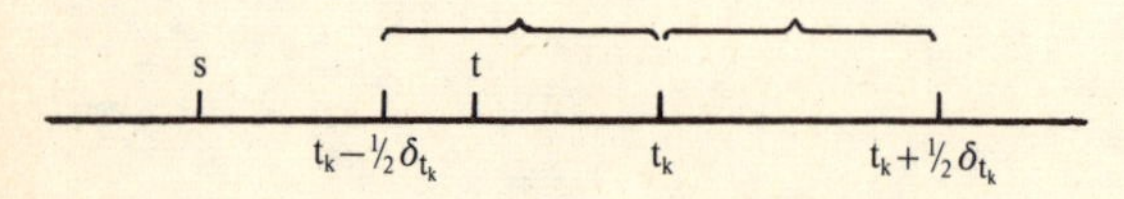

Fig. 2.1. An element of covering (2.6) containing t □

Theorem 2.6. Let X and $Y : I \to L_2(\Omega)$ be m.s. continuous on I. If $f : I \to \mathbb{R}$ is continuous on I and $a, b \in \mathbb{R}$, then $aX(t) + bY(t)$ and $f(t)X(t)$, $t \in I$, are m.s. continuous on I, and $\|X(t)\|$ and $\mathrm{E}X(t)$, $t \in I$, are continuous on I.

Proof. These assertions are immediate consequences of Theorems 2.3, 4, and of the following inequalities:

$$| \|X(t')\| - \|X(t)\| | \leq \|X(t') - X(t)\| \tag{2.7}$$

and, see (1.67),

$$|\mathrm{E}X(t') - \mathrm{E}X(t)| \leq \|X(t') - X(t)\| . \tag{2.8}$$ □

A number of m.s. continuity properties of a second-order stochastic process are tied to the properties of its correlation function, as illustrated in the following theorem.

Theorem 2.7. (a) $X : I \to L_2(\Omega)$ is m.s. continuous at $t \in I$ if and only if $\mathrm{E}\{X(t_1)X(t_2)\}$ is continuous at (t, t).

b) Furthermore, $X : I \to L_2(\Omega)$ is m.s. continuous on I if and only if $\mathrm{E}\{X(t_1)X(t_2)\}$ is continuous on I^2.

Proof. The "only if" parts of (a) and (b) follow from the continuity of the inner product property discussed in Sect. 2.1. The "if" parts are true since

$$\mathrm{E}\{X(s) - X(t)\}^2 = \mathrm{E}X^2(s) - 2\mathrm{E}\{X(s)X(t)\} + \mathrm{E}X^2(t) \to 0 \quad \text{as} \quad s \to t .$$ □

As shown in Sect. 1.4.4, an example of a m.s. continuous second-order process is the Wiener-Lévy process $W : [0, T] \to L_2(\Omega)$ since, for each $t \in [0, T]$,

$$\|W(s) - W(t)\| \to 0 \quad \text{as} \quad s \to t \text{ in } [0, T] .$$

We have also seen in Sect. 1.4.4 that

$$\mathrm{E}W(t) = 0, \ \|W(t)\| = \sqrt{t} , \quad t \in [0, T] \quad \text{and}$$

$$\mathrm{E}\{W(s)W(t)\} = \min(s, t) , \quad (s, t) \in [0, T]^2 .$$

Thus, $\mathrm{E}W(t)$ and $\|W(t)\|$ are continuous on $[0, T]$ and $\mathrm{E}\{W(s)W(t)\}$ is continuous on $[0, T]^2$.

2.3 Differentiability in Mean Square

As before, let I be an interval of the real line and let $X : I \to L_2(\Omega)$ be a second-order stochastic process whose random variables are $X(t)$, $t \in I$.

Definition. (a) X is *differentiable in mean square* (m.s. *differentiable*) *at* $t \in I$ if

$$\underset{s \to t}{\text{l.i.m.}} \frac{X(s) - X(t)}{s - t}$$

exists. This limit will be denoted by $dX(t)/dt$, $DX(t)$, or $X'(t)$. Since limits in m.s. belong to $L_2(\Omega)$, $X'(t) \in L_2(\Omega)$.

b) Furthermore X *is* m.s. *differentiable on* I if it is m.s. differentiable at each $t \in I$. The m.s. derivative is then the mapping $X' : I \to L_2(\Omega)$.

The following theorem is an immediate consequence of Theorem 2.2.

Theorem 2.8. At $t \in I$ X is m.s. differentiable if and only if, for each sequence $\{s_n\}_{n \in N}$ in I converging to t as $n \to \infty$,

$$\left\{ \frac{X(s_n) - X(t)}{s_n - t} \right\}_{n \in N}$$

is a Cauchy sequence in $L_2(\Omega)$. The (unique) m.s. limit of these Cauchy sequences is the m.s. derivative $X'(t)$.

The following criterion for m.s. differentiability follows directly from the m.s. convergence criterion [see (C) in Sect. 2.1].

Theorem 2.9. At $t \in I$ X is m.s. differentiable if and only if, with $t + h$ and $t + k$ belonging to $I \backslash \{t\}$,

$$\frac{1}{hk} \underset{h}{\Delta} \underset{k}{\Delta} \mathrm{E}\{X(t) X(t)\} = \mathrm{E}\left\{ \left(\frac{X(t+h) - X(t)}{h} \right) \left(\frac{X(t+k) - X(t)}{k} \right) \right\}$$

converges as $h, k \to 0$ independently.

As an example, we shall show that the Wiener-Lévy process $W : [0, T] \to L_2(\Omega)$ is nowhere m.s. differentiable. Applying the above criterion with $h = k$, we have

$$\mathrm{E}\left\{ \left(\frac{W(t+h) - W(t)}{h} \right) \left(\frac{X(t+h) - W(t)}{h} \right) \right\}$$

$$= \frac{1}{h^2} \mathrm{E}\{W(t+h) - W(t)\}^2 = \frac{1}{h}$$

which does not converge as $h \to 0$.

Theorem 2.10. Differentiability in mean square of $X : I \to L_2(\Omega)$ at $t \in I$ implies m.s. continuity of X at t.

Proof. We first note that according to (A) in Sect. 2.1,

$$\lim_{h\to 0} \left\| \frac{X(t+h)-X(t)}{h} \right\| = \left\| \underset{h\to 0}{\text{l.i.m.}} \frac{X(t+h)-X(t)}{h} \right\| .$$

Hence, with $t + h \in I$,

$$\|X(t+h) - X(t)\| = |h| \left\| \frac{X(t+h)-X(t)}{h} \right\| \to (0)\, \|X'(t)\| = 0$$

as $h \to 0$. □

Theorem 2.11. If X is m.s. differentiable at $t \in I$, then $\mathrm{E}X : I \to \mathbb{R}$ is differentiable at t and

$$\frac{d}{dt}\mathrm{E}X(t) = \mathrm{E}X'(t) . \tag{2.9}$$

Proof. Following (A) of Sect. 2.1, we see that

$$\left\{ \frac{X(t+h)-X(t)}{h} \xrightarrow{\text{m.s.}} X'(t) \quad \text{as} \quad h \to 0 \right\}$$

implies

$$\mathrm{E}\left\{ \frac{X(t+h)-X(t)}{h} \right\} \to \mathrm{E}X'(t) , \quad \text{i.e.,}$$

$$\frac{\mathrm{E}X(t+h)-\mathrm{E}X(t)}{h} \to \mathrm{E}X'(t) .$$

Thus we have (2.9). □

Let both X and Y be mappings of I into $L_2(\Omega)$ whose random variables are, respectively, $X(t)$ and $Y(t)$, $t \in I$.

Theorem 2.12. If X and Y are m.s. differentiable at $t \in I$ and if a and b are real numbers, then $aX + bY$ is m.s. differentiable with

$$\frac{d}{dt}[aX(t) + bY(t)] = aX'(t) + bY'(t) . \tag{2.10}$$

Proof. Equation (2.10) follows from Theorem 2.3 by replacing $X(s)$ by $[X(s) - X(t)]/(s - t)$ and $Y(s)$ by $[Y(s) - Y(t)]/(s - t)$. □

Theorem 2.13. If $X : I \to L_2(\Omega)$, then $X_o = X - \mathrm{E}X$ is a centered stochastic process. If we write $X = X_o + \mathrm{E}X$ and if X is m.s. differentiable at $t \in I$, then $d/dt\, X_o(t)$ and $d/dt\, \mathrm{E}X(t)$ exist and

$$X'(t) = X_o'(t) + \frac{d}{dt}\mathrm{E}X(t) . \tag{2.11}$$

Proof. This result follows from Theorems 2.11, 12 by treating EX as a degenerate stochastic process although it is a deterministic function. □

Theorem 2.14. If X is m.s. differentiable at $s \in I$ and Y at $t \in I$, then

$$\mathrm{E}\{X'(s)Y(t)\} = \frac{\partial}{\partial s}\mathrm{E}\{X(s)Y(t)\} \tag{2.12}$$

$$\mathrm{E}\{X(s)Y'(t)\} = \frac{\partial}{\partial t}\mathrm{E}\{X(s)Y(t)\} \tag{2.13}$$

$$\mathrm{E}\{X'(s)Y'(t)\} = \frac{\partial^2}{\partial s\,\partial t}\mathrm{E}\{X(s)Y(t)\} . \tag{2.14}$$

In particular, these results are valid for $X = Y$.

Proof. Equations (2.12, 13) follow from (A) of Sect 2.1 since

$$\begin{aligned}
\mathrm{E}\{X'(s)Y(t)\} &= \mathrm{E}\left\{\underset{h\to 0}{\mathrm{l.i.m.}}\left(\frac{X(s+h)-X(s)}{h}\right)Y(t)\right\} \\
&= \lim_{h\to 0}\mathrm{E}\left\{\left(\frac{X(s+h)-X(s)}{h}\right)Y(t)\right\} \\
&= \lim_{h\to 0}\frac{\mathrm{E}\{X(\mathrm{s}+\mathrm{h})Y(t)\}-\mathrm{E}\{X(s)Y(t)\}}{h} \\
&= \frac{\partial}{\partial s}\mathrm{E}\{X(s)Y(t)\} .
\end{aligned}$$

For (2.14), the above result together with (A) of Sect. 2.1 leads to

$$\begin{aligned}
\mathrm{E}\{X'(s)Y'(t)\} &= \mathrm{E}\left\{\underset{h\to 0}{\mathrm{l.i.m.}}\left(\frac{X(s+h)-X(s)}{h}\right)Y'(t)\right\} \\
&= \frac{\partial}{\partial s}\mathrm{E}\left\{X(s)\,\underset{k\to 0}{\mathrm{l.i.m.}}\left(\frac{Y(t+k)-Y(t)}{k}\right)\right\} \\
&= \frac{\partial}{\partial s}\left\{\lim_{k\to 0}\frac{\mathrm{E}\{X(s)Y(t+k)\}-\mathrm{E}\{X(s)Y(t)\}}{k}\right\} \\
&= \frac{\partial^2}{\partial s\,\partial t}\mathrm{E}\{X(s)Y(t)\} .
\end{aligned}$$

□

Theorem 2.15. Let $I = [a, b] \in \mathbb{R}$. If $X : I \to L_2(\Omega)$ is m.s. differentiable on I, then

$$\forall t \in I : X'(t) = 0$$

if and only if X is a constant random function. Then $X(s) = X(t)$ a.s. for all s and t in I.

Proof. For the "if" part, let $Z : \Omega \to \mathbb{R}$ be the second-order random variable such that

$$\forall t \in I, \quad X(t) = Z .$$

Then,

$$\forall t \in I : \frac{X(t+h) - X(t)}{h} = \frac{Z - Z}{h} = 0$$

and hence, $X'(t) = 0$.

Consider the "only if" part. Since $\forall t \in I : X'(t) = 0$, it follows from Theorem 2.14 that

$$\forall (s, t) \in I^2 : \frac{\partial}{\partial s} \mathrm{E}\{X(s)X(t)\} = 0 \quad \text{and} \quad \frac{\partial}{\partial t} \mathrm{E}\{X(s)X(t)\} = 0 .$$

Thus, $\mathrm{E}\{X(s)X(t)\} = c$, a constant, on I^2. Hence, for any $a \in I$,

$$\begin{aligned} \forall t \in I : \|X(t) - X(a)\|^2 &= \mathrm{E}X^2(t) - 2\mathrm{E}\{X(t)X(a)\} + \mathrm{E}X^2(a) \\ &= c - 2c + c = 0 \end{aligned}$$

and we have $X(t) = X(a)$. □

2.3.1 Supplementary Exercises

Exercise 2.1. Show that, if I is an interval of $\mathbb{R}$, if $X : I \to L_2(\Omega)$ is m.s. differentiable at $t \in I$, and if $f : I \to \mathbb{R}$ is differentiable at t, then fX is m.s. differentiable at t and

$$\frac{d}{dt} f(t) X(t) = f'(t) X(t) + f(t) X'(t) . \tag{2.15}$$

Exercise 2.2. Given an orthonormal sequence of random variables

$$\{Z_n\}_{n \in \mathbb{N}} : \mathrm{E}\{Z_i Z_j\} = \delta_{ij} , \quad i, j \in \mathbb{N} ,$$

and let a second-order process $X : [0, 1] \to L_2(\Omega)$ be defined as follows:

$$X(0) = 0$$

$$X(t) = Z_k\,, \quad \frac{1}{2^k} < t \leq \frac{1}{2^{k-1}}\,, \quad k \in \mathbb{N}\,.$$

a) Compute $\mathrm{E}\{X(s)X(t)\}$, $(s, t) \in [0, 1]^2$.
b) Show that

$$\frac{\partial^2}{\partial s\,\partial t}\mathrm{E}\{X(s)X(t)\}$$

exists at (0, 0) and is equal to 0.
c) Show that X is not m.s. differentiable at 0 by observing that

$$\frac{1}{hk}\underset{h}{\Delta}\underset{k}{\Delta}\,\mathrm{E}\{X(0)X(0)\}$$

does not converge as $h, k \to 0$ independently.

2.4 Integration in Mean Square

Since in our discourse we need only certain integrals of the Riemann-Stieltjes type, our discussions are restricted to stochastic integrals of this type. Some integrals of a slightly different kind will be treated in Chap. 4.

Let $I = [a, b] \subset \mathbb{R}$, $f : I \to \mathbb{R}$, and $X : I \to L_2(\Omega)$. We consider in what follows the Riemann-Stieltjes integrals of the forms

$$\int_a^b f(t)\,dX(t) \quad \text{and} \quad \int_a^b X(t)\,df(t)$$

in the mean-square sense.

First, some preliminaries are in order. Let $t_o, t_1, \ldots, t_k$ be points of $[a, b]$ such that

$$a = t_o < t_1 < \ldots < t_k = b$$

and let t_i' be an arbitrary point in $[t_{i-1}, t_i]$, $i = 1, \ldots, k$. The set

$$p = \{[t_{i-1}, t_i], i = 1, \ldots, k\} \cup \{t_i', i = 1, \ldots, k\} \tag{2.16}$$

is called a *partition* of $[a, b]$. The set of all partitions of $[a, b]$ will be denoted by $P[a, b]$.

In the above, the t_i are called *subdivision points* and t_i' *intermediate points*, which are incorporated into the partitions to formulate what follows efficiently. Hence, there are infinitely many partitions of $[a, b]$ with the same subdivision points.

A partition p' is called a *refinement* of p if each subdivision point of p is also a subdivision of p'. The quantity

$$\Delta p = \max_{i=1,\ldots,k} (t_i - t_{i-1}) \tag{2.17}$$

is called the *mesh* of p. A sequence $\{p_n\}_{n\in N}$ of partitions of $[a, b]$ is called *convergent* if $\Delta p_n \to 0$ as $n \to \infty$.

Given f, X, and the partition p of $[a, b]$, we define the *Riemann-Stieltjes* (R–S) *sums* by

$$S_{f,X}(p) = \sum_{i=1}^{k} f(t_i')[X(t_i) - X(t_{i-1})] \quad \text{and} \tag{2.18}$$

$$S_{X,f}(p) = \sum_{i=1}^{k} X(t_i')[f(t_i) - f(t_{i-1})] \,. \tag{2.19}$$

They are elements of $L_2(\Omega)$.

Definition. If for each convergent sequence of partitions $\{p_n\}_{n\in N}$ of $[a, b]$ the sequence $\{S_{f,X}(p_n)\}_{n\in N}$ is a Cauchy sequence in $L_2(\Omega)$, then f is said to be *m.s. R–S integrable* on $[a, b]$ with respect to X. The m.s. limit of this Cauchy sequence is called the *R–S m.s. integral* on $[a, b]$ of f with respect to X, and is denoted by

$$\int_a^b f(t)\, dX(t) \,. \tag{2.20}$$

Definition. If for each convergent sequence of partitions $\{p_n\}_{n\in N}$ of $[a, b]$ the sequence $\{S_{X,f}(p_n)\}_{n\in N}$ is a Cauchy sequence in $L_2(\Omega)$, then X is said to be *m.s. R–S integrable* on $[a, b]$ with respect to f. The m.s. limit of this Cauchy sequence is called the *R–S m.s. integral* on $[a, b]$ of X with respect to f, and is denoted by

$$\int_a^b X(t)\, df(t) \,. \tag{2.21}$$

It is seen that the integrals defined in (2.20, 21) are elements of $L_2(\Omega)$. Part (a) of the following theorem is a consequence of Theorem 2.2.

Theorem 2.16. (a) The definitions given above are admissible, i.e., all Cauchy sequences involved in the definitions have one and the same limit in $L_2(\Omega)$.

b) Furthermore, f, (X) is R–S m.s. integrable on $[a, b]$ with respect to X, (f) if and only if there is an element in $L_2(\Omega)$, denoted by

$$\int_a^b f(t)\, dX(t) \,, \qquad \left(\int_a^b X(t)\, df(t)\right)$$

such that, for each $\varepsilon > 0$, there is a $\delta > 0$ with the property that

$$\Delta p < \delta \Rightarrow \left\| \int_a^b f(t)\, dX(t) - S_{f,X}(p) \right\| < \varepsilon ,$$

$$\left(\left\| \int_a^b X(t)\, df(t) - S_{X,f}(p) \right\| < \varepsilon \right) .$$

We observe that the positions of the intermediate points take no part in the foregoing theorem and definitions.

Definition. If

$$\int_a^b f(t)\, dX(t) , \qquad \left[\int_a^b X(t)\, df(t) \right]$$

exists, then

$$\int_b^a f(t)\, dX(t) = - \int_a^b f(t)\, dX(t) , \qquad \left[\int_b^a X(t)\, df(t) = - \int_a^b X(t)\, df(t) \right]$$

and hence

$$\int_a^a f(t)\, dX(t) = 0 , \qquad \left[\int_a^a X(t)\, df(t) = 0 \right] .$$

In the special case where $\forall t \in [a, b]$, $f(t) = t$, the R–S sum defined in (2.19) reduces to a *Riemann sum of X* in the form

$$S_{X,f} = \sum_{i=1}^{k} X(t_i')(t_i - t_{i-1}) \tag{2.22}$$

and (2.21) reduces to the *m.s. Riemann integral*

$$\int_a^b X(t)\, dt . \tag{2.23}$$

2.4.1 Some Elementary Properties

Let $I = [a, b] \subset \mathbb{R}$, $f : I \to \mathbb{R}$ and $X : \mathrm{I} \to \mathrm{L}_2(\Omega)$.

Theorem 2.17. *(Partial Integration).* The m.s. integral

$$\int_a^b f(t)\, dX(t)$$

exists if and only if the m.s. integral

$$\int_a^b X(t)\, df(t)$$

exists and

$$\int_a^b f(t)\, dX(t) = f(t)X(t) \Big|_a^b - \int_a^b X(t)\, df(t) . \tag{2.24}$$

Proof. Consider partition p as defined in (2.16) and define partition p' of $[a, b]$ by

$$p' = \{[t_i', t_{i+1}'], i = 0, \ldots, k\} \cup \{t_i, i = 0, \ldots, k\}$$

with $t_o' = a$ and $t_{k+1}' = b$. In the above, t_i' is a subdivision point whereas t_i is an intermediate point in $[t_i', t_{i+1}']$. Let $S_{f,X}(p)$ and $S_{X,f}(p)$ denote the R–S sums as defined in (2.18, 19). It is seen that

$$\begin{aligned} S_{f,X}(p) &= \sum_{i=1}^{k} f(t_i')[X(t_i) - X(t_{i-1})] = \sum_{i=1}^{k} f(t_i')X(t_i) - \sum_{i=0}^{k-1} f(t_{i+1}')X(t_i) \\ &= -\sum_{i=0}^{k} X(t_i)[f(t_{i+1}') - f(t_i')] + X(t_k)f(t_{k+1}') - X(t_o)f(t_o') \\ &= -S_{X,f}(p') + f(b)X(b) - f(a)X(a) \, . \end{aligned} \tag{2.25}$$

Now, suppose

$$\int_a^b X(t)\, df(t)$$

exists and let $\{p_n\}_{n\in\mathbb{N}}$ be an arbitrary convergent sequence of partitions of $[a, b]$. For each p_n, define the partition p_n' of $[a, b]$ in the same manner as p' to p. Then $\{p_n'\}_{n\in\mathbb{N}}$ is also a convergent sequence of partitions of $[a, b]$. Hence, $\{S_{X,f}(p_n')\}_{n\in\mathbb{N}}$ is a Cauchy sequence in $L_2(\Omega)$, and so is $\{S_{f,X}(p_n)\}_{n\in\mathbb{N}}$, as seen from (2.25). The integral

$$\int_a^b f(t)\, dX(t)$$

thus exists and (2.24) is proved.

The proof follows analogously when the existence of

$$\int_a^b f(t)\, dX(t)$$

is given. □

Theorem 2.18. If $a \leqslant c \leqslant b$ and if

$$\int_a^b f(t)\, dX(t) \, , \quad \int_a^c f(t)\, dX(t) \, , \quad \text{and} \quad \int_c^b f(t)\, dX(t)$$

exist, then

$$\int_a^b f(t)\, dX(t) = \int_a^c f(t)\, dX(t) + \int_c^b f(t)\, dX(t) \quad \text{and} \tag{2.26}$$

$$\int_a^b X(t)\, df(t) = \int_a^c X(t)\, df(t) + \int_c^b X(t)\, df(t) \, . \tag{2.27}$$

Proof. Let

$$\{p_n[a, c]\}_{n\in N} \quad \text{and} \quad \{p_n[c, b]\}_{n\in N}$$

be convergent sequences of partitions of $[a, c]$ and $[c, b]$, respectively. Then,

$$\{p_n[a, b]\}_{n\in N} = \{p_n[a, c] \cup p_n[c, b]\}_{n\in N}$$

is a convergent sequence of partitions of $[a, b]$. For the corresponding R–S sums we obtain

$$S_{f,X}(p_n[a, c]) + S_{f,X}(p_n[c, b]) = S_{f,X}(p_n[a, b])$$

whose members converge in m.s., as $n \to \infty$, to the corresponding integrals in (2.26). Equation (2.27) is proved analogously. □

We shall omit the easy proof of the next theorem.

Theorem 2.19. If X and $Y : [a, b] \to L_2(\Omega)$ and f and $g : [a, b] \to \mathbb{R}$ are such that

$$\int_a^b f(t)\, dX(t)\,, \qquad \int_a^b g(t)\, dX(t)\,, \quad \text{and} \quad \int_a^b f(t)\, dY(t)$$

exist, then, if p and q are real numbers, all m.s. integrals given below exist and

$$\int_a^b [pf(t) + qg(t)]\, dX(t) = p \int_a^b f(t)\, dX(t) + q \int_a^b g(t)\, dX(t) \tag{2.28}$$

$$\int_a^b f(t)\, d[pX(t) + qY(t)] = p \int_a^b f(t)\, dX(t) + q \int_a^b f(t)\, dY(t)\,. \tag{2.29}$$

Similar equalities hold for integrals of the type

$$\int_a^b X(t)\, df(t)\,.$$

Theorem 2.20. If

$$\int_a^b f(t)\, dX(t) \quad \text{or} \quad \int_a^b X(t)\, df(t)$$

exists, then

$$\int_a^b f(t)\, d\,\mathrm{E}X(t) \quad \text{and} \quad \int_a^b \mathrm{E}X(t)\, df(t)$$

exist and

$$\mathrm{E}\left\{\int_a^b f(t)\,dX(t)\right\} = \int_a^b f(t)\,d\,\mathrm{E}X(t) \tag{2.30}$$

$$\mathrm{E}\left\{\int_a^b X(t)\,df(t)\right\} = \int_a^b \mathrm{E}X(t)\,df(t)\,. \tag{2.31}$$

Proof. Let p be the partition of $[a, b]$ defined in (2.16). Then

$$S_{f,X}(p) = \sum_{i=1}^{k} f(t_i')[X(t_i) - X(t_{i-1})] \quad \text{and}$$

$$\mathrm{E}\{S_{f,X}(p)\} = \sum_{i=1}^{k} f(t_i')[\mathrm{E}X(t_i) - \mathrm{E}X(t_{i-1})]\,. \tag{2.32}$$

The right-hand side of (2.32) is a R–S sum corresponding to the ordinary R–S integral

$$\int_a^b f(t)\,d\,\mathrm{E}X(t)\,.$$

Now, let $\{p_n\}_{n\in\mathbb{N}}$ be a convergent sequence of partitions of $[a, b]$. Then, as $n \to \infty$,

$$S_{f,X}(p_n) \xrightarrow{\text{m.s.}} \int_a^b f(t)\,dX(t)$$

and hence, as seen from (A) of Sect. 2.1,

$$\mathrm{E}\{S_{f,X}(p_n)\} \to \mathrm{E}\left\{\int_a^b f(t)\,dX(t)\right\}. \tag{2.33}$$

Thus, $\{\mathrm{E}S_{f,X}(p_n)\}_{n\in\mathbb{N}}$ is a convergent sequence in $\mathbb{R}$ and, in view of (2.32),

$$\mathrm{E}\{S_{f,X}(p_n)\} \to \int_a^b f(t)\,d\,\mathrm{E}X(t) \tag{2.34}$$

as $n \to \infty$, showing the existence of the latter integral.

The first part of the theorem follows from (2.33, 34). The second part follows analogously. □

Let $X_o = X - \mathrm{E}X$ be a centered process and view $\mathrm{E}X$ as a degenerate random process. Theorems 2.19, 20 lead to the following result.

Theorem 2.21. If

$$\int_a^b f(t)\,dX(t) \quad \text{or} \quad \int_a^b X(t)\,df(t)$$

exists, then

$$\int_a^b f(t)\,dX_o(t)\,, \quad \int_a^b f(t)\,d\mathrm{E}X(t)\,, \quad \int_a^b X_o(t)\,df(t)\,, \quad \text{and} \quad \int_a^b \mathrm{E}X(t)\,df(t)$$

exist and

$$\int_a^b f(t)\,dX(t) = \int_a^b f(t)\,dX_o(t) + \int_a^b f(t)\,d\mathrm{E}X(t) \tag{2.35}$$

$$\int_a^b X(t)\,df(t) = \int_a^b X_o(t)\,df(t) + \int_a^b \mathrm{E}X(t)\,df(t)\,. \tag{2.36}$$

2.4.2 A Condition for Existence

Let $I = [a, b] \subset \mathbb{R}, f: I \to \mathbb{R}$, and $X: I \to L_2(\Omega)$. Let p be a partition of I and $P = \{p\}$ the set of all partitions of I.

Definition. *The variation of* f on I with respect to p is defined as

$$V_f(p) = \sum_{i=1}^{k} |f(t_i) - f(t_{i-1})| \tag{2.37}$$

and the *total variation of* f on I as

$$V_f(I) = \sup_{p \in P} V_f(p)\,. \tag{2.38}$$

The function f is said to be of bounded variation on I as $V_f(I)$ is finite.

We recall that monotonic functions and differentiable functions with bounded derivatives on I are of bounded variation on I. In the first case,

$$V_f(I) = |f(b) - f(a)|$$

and, in the second,

$$V_f(I) \leqslant c\,(b - a)$$

if $|f'(t)| \leqslant c$ for all $t \in I$.

Furthermore, if p' is a refinement of p, then

$$V_f(p') \geqslant V_f(p)$$

and, if $a \leqslant c \leqslant b$ and f is of bounded variation on $[a, b]$,

$$V_f([a, b]) = V_f([a, c]) + V_f([c, b])\,.$$

Theorem 2.22. If X is m.s. continuous and f is of bounded variation on I, then

a) $\int_a^b X(t)\,df(t)$ exists $\left(\text{and hence also } \int_a^b f(t)\,dX(t)\right)$;

b) $\left\| \int_a^b X(t)\, df(t) \right\| \leqslant MV_f(I)\,, \qquad \text{where} \quad M = \max_{t \in I} \|X(t)\|\,;$

c) $\int_a^b X(t)\, dt$ and $\int_a^b \|X(t)\|\, dt$ also exist and

$$\left\| \int_a^b X(t)\, dt \right\| \leqslant \int_a^b \|X(t)\|\, dt \leqslant M(b-a)\,;$$

d) and if X^* is also m.s. continuous on I,

$$\mathrm{E}\left\{ \int_a^b X(s)\, ds \int_a^b X^*(t)\, dt \right\} = \iint_{[a,b]^2} \mathrm{E}\{X(s) X^*(t)\}\, ds\, dt\,.$$

e) The m.s. integrals given above have all the properties discussed in Sect. 2.4.1.

Proof. Let $\{p_n\}_{n \in N}$ be a converegent sequence of partitions of I. To prove (a), it is sufficient to show that (Theorem 2.16)

$$\{S_{X,f}(p_n)\}_{n \in N}$$

is a Cauchy sequence in $L_2(\Omega)$. Let $\varepsilon > 0$. Because of its m.s. continuity, X is uniformly m.s. continuous on the compact set I by virtue of Theorem 2.5. Hence, there is a number $\delta(\varepsilon) > 0$ such that

$$|s-t| < \delta \Rightarrow \|X(s) - X(t)\| < \varepsilon,\ s \text{ and } t \text{ in } I\,.$$

Since $\Delta p_n \to 0$ as $n \to \infty$, there is a number $N(\delta(\varepsilon))$ such that

$$n > N \Rightarrow \Delta p_n < \delta\,.$$

Now, let $m, n > N$ and let p be a partition of I, refining both p_m and p_n. We have

$$\|S_{X,f}(p_m) - S_{X,f}(p_n)\| \leqslant \|S_{X,f}(p_m) - S_{X,f}(p)\| + \|S_{X,f}(p) - S_{X,f}(p_n)\|\,. \tag{2.39}$$

Consider the first term on the right-hand side and let $[t_{k-1}, t_k]$ be an interval of the partition p_m and t'_k its intermediate point. Since p refines p_m, t_{k-1} and t_k are also subdivision points of p. Let $s_{k,o}, s_{k,1}, \ldots, s_{k,j}$ be subdivision points of p belonging to the interval $[t_{k-1}, t_k]$ of p_m with

$$t_{k-1} = s_{k,o} < s_{k,1} < \ldots < s_{k,j} = t_k\,,$$

and let $s'_{k,1}, \ldots, s'_{k,j}$ be the corresponding intermediate points of p (Fig. 2.2).

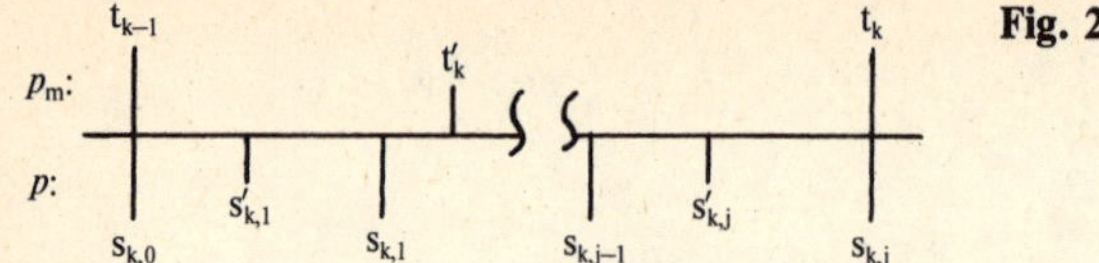

Fig. 2.2. Partitions p_m and p

From

$$A_k = X(t'_k)[f(t_k) - f(t_{k-1})] - \sum_{i=1}^{j} X(s'_{k,i})[f(s_{k,i}) - f(s_{k,i-1})] \tag{2.40}$$

it is seen that

$$S_{X,f}(p_m) - S_{X,f}(p) = \sum_k A_k \,. \tag{2.41}$$

Since

$$f(t_k) - f(t_{k-1}) = \sum_{i=1}^{j} [f(s_{k,i}) - f(s_{k,i-1})] \,,$$

(2.40) becomes

$$A_k = \sum_{i=1}^{j} [X(t'_k) - X(s'_{k,i})][f(s_{k,i}) - f(s_{k,i-1})] \,.$$

And since $\Delta p_m < \delta$, it follows that

$$\|X(t'_k) - X(s'_{k,i})\| < \varepsilon \,,$$

and hence

$$\begin{aligned} \|A_k\| &\leqslant \sum_{i=1}^{j} \|X(t'_k) - X(s'_{k,i})\| \; |f(s_{k,i}) - f(s_{k,i-1})| \\ &\leqslant \sum_{i=1}^{j} \varepsilon \, |f(s_{k,i}) - f(s_{k,i-1})| \leqslant \varepsilon \, V_f([t_{k-1}, t_k]) \,. \end{aligned}$$

Finally, (2.41) leads to

$$\begin{aligned} \|S_{X,f}(p_m) - S_{X,f}(p)\| &\leqslant \sum_k \|A_k\| \leqslant \varepsilon \sum_k V_f([t_{k-1}, t_k]) \\ &= \varepsilon \, V_f\Big(\bigcup_k [t_{k-1}, t_k]\Big) = \varepsilon \, V_f(I) \,. \end{aligned} \tag{2.42}$$

Reverting to (2.39), it follows from (2.42) that

$$\|S_{X,f}(p) - S_{X,f}(p_n)\| \leqslant \varepsilon \, V_f(I) \quad \text{and}$$

$$\|S_{X,f}(p_m) - S_{X,f}(p_n)\| \leqslant 2\varepsilon \, V_f(I)$$

if $m, n > N$. The proof of (a) is complete since $V_f(I)$ is a finite constant.

For (b), let us first remark that $\|X(t)\|$ is continuous on the compact set I. Hence, M exists. The inequality in (b) follows, since for any partition p of I

$$\begin{aligned}\|S_{X,f}(p)\| &= \left\| \sum_k X(t_k')[f(t_k)-f(t_{k-1})] \right\| \\ &\leqslant \sum_k \|X(t_k')\| \, |f(t_k)-f(t_{k-1})| \\ &\leqslant M \sum_k |f(t_k)-f(t_{k-1})| \leqslant MV_f(I) \,. \end{aligned} \tag{2.43}$$

As for (c), the function $f(t) = t$, $t \in I$, is of bounded variation on I with total variation $(b-a)$. Hence,

$$\int_a^b X(t)\, dt$$

exists as X is m.s. continuous on I. The inequalities in (c) follow from [see (2.43)]

$$\|S_{X,f}(p_n)\| \leqslant M \sum_k |t_k - t_{k-1}| = M(b-a)$$

by letting $n \to \infty$ if $\{p_n\}_{n \in N}$ is a convergent sequence of partitions of I. Here we have used the result

$$\sum_k \|X(t_k')\| \; |t_k - t_{k-1}| \to \int_a^b \|X(t)\|\, dt$$

as this integral exists due to the continuity of $\|X(t)\|$ on I.

Concerning (d),

$$\int_a^b X(s)\, ds \quad \text{and} \quad \int_a^b X^*(t)\, dt$$

exist as Riemann m.s. integrals and, due to continuity of $\mathrm{E}\{X(s)X^*(t)\}$ on $[a,b]^2$,

$$\iint_{[a,b]^2} \mathrm{E}\{X(s)X^*(t)\}\, ds\, dt$$

exists as an ordinary Riemann integral.

If

$$S_m = \sum_j X(s_j')(s_j - s_{j-1}) \quad \text{and}$$

$$S_n^* = \sum_k X^*(t_k')(t_k - t_{k-1}) \,, \quad \text{then}$$

$$\mathrm{E}\left\{\int_a^b X(s)\, ds \int_a^b X^*(t)\, dt\right\} = \mathrm{E}\left\{\lim_{m\to\infty} S_m \lim_{n\to\infty} S_n^*\right\} = \lim_{m,n\to\infty} \mathrm{E}\{S_m S_n^*\}$$

on account of the continuity of the inner product.

On the other hand,

$$\begin{aligned} E\{S_m S_n^*\} &= \sum_j \sum_k E\{X(s_j')X^*(t_k')(s_j - s_{j-1})(t_k - t_{k-1})\} \\ &\to \iint_{[a,b]^2} E\{X(s)X^*(t)\}\, ds\, dt \end{aligned}$$

as $n, m \to \infty$ according to the definition of the Riemann integral of a real function. □

Theorem 2.23. If X is m.s. continuous and f continuously differentiable on I, then the integrals given below exist and are a.s. identical:

$$\int_a^b X(t)\, df(t) = \int_a^b X(t) f'(t)\, dt\,. \tag{2.44}$$

Proof. Continuous differentiability of f on I implies bounded variation on I. Hence, the first integral in (2.44) exists according to (a) of Theorem 2.22.

Since f' is continuous and X m.s. continuous on I, Xf' is m.s. continuous on I and hence the second integral exists by virtue of (b) of Theorem 2.22.

Now, to show equality of the integrals, we may use the same convergent sequence $\{p_n\}_{n\in\mathbb{N}}$ of partitions of I to both integrals. Let p_n be the partition p defined in (2.16). Then, $S_{X,f}(p_n)$ is the R–S sum corresponding to the first integral. Since f is continuously differentiable, for each i there is a number t_i'' in $[t_{i-1}, t_i]$ such that

$$f(t_i) - f(t_{i-1}) = f'(t_i'')(t_i - t_{i-1})\,.$$

As the subdivision points in the partitions may be chosen arbitrarily in their intervals, we set

$$t_i' = t_i''$$

for all i and the R–S sum $S_{X,f}(p_n)$ becomes

$$\begin{aligned} S_{X,f}(p_n) &= \sum_{i=1}^{k} X(t_i')[f(t_i) - f(t_{i-1})] \\ &= \sum_{i=1}^{k} X(t_i'') f'(t_i'')(t_i - t_{i-1})\,. \end{aligned}$$

This is clearly a Riemann sum $S_{Xf'}(p_n)$ corresponding to the second integral. Hence, by a suitable choice of the subdivision points in the partitions p_n, for each n we have

$$S_{X,f}(p_n) = S_{Xf'}(p_n)\,.$$

Equation (2.44) follows as $n \to \infty$. □

Theorem 2.24. Let $X : I \to L_2(\Omega)$ be m.s. continuous on $I = [a, b]$. Then,

$$\int_a^t X(s)\,ds$$

is m.s. differentiable (and hence m.s. continuous) on I and

$$\frac{d}{dt}\int_a^t X(s)\,ds = X(t),\; t \in I\,. \tag{2.45}$$

Proof. Suppose t and $t + h$, $h > 0$, are in I. We have

$$\left\| \frac{1}{h}\left[\int_a^{t+h} X(s)\,ds - \int_a^t X(s)\,ds\right] - X(t) \right\|$$

$$= \left\| \frac{1}{h}\int_t^{t+h} X(s)\,ds - \frac{1}{h}X(t)\int_t^{t+h} ds \right\|$$

$$= \frac{1}{h}\left\| \int_t^{t+h} [X(s) - X(t)]\,ds \right\|$$

$$\leqslant \frac{1}{h}h \max_{s\in[t,t+h]} \|X(s) - X(t)\| \to 0$$

as $h \to 0$. The inequality is an application of Theorem 2.22. □

Theorem 2.25. If $X : I \to L_2(\Omega)$ is such that $X'(t)$ is m.s. continuously differentiable on $I = [a, b]$, then

$$\int_a^b X'(t)\,dt = X(b) - X(a)\,. \tag{2.46}$$

Proof. Define the mapping $Y : I \to L_2(\Omega)$ as

$$Y(t) = \int_a^t X'(s)\,ds - X(t)$$

for all $t \in I$. Then

$$Y'(t) = X'(t) - X'(t) = 0$$

for all $t \in I$ and hence $Y(t)$ is a.s. one and the same element of $L_2(\Omega)$ for all $t \in I$ (Theorem 2.15). So $Y(a) = Y(t)$ for all $t \in I$, and we have

$$\int_a^a X'(s)\,ds - X(a) = \int_a^t X'(s)\,ds - X(t)\,, \quad t \in I\,.$$

Equation (2.46) follows by letting $t = b$ since

$$\int_a^a X'(s)\,ds = 0\,.$$

□

2.4.3 A Strong Condition for Existence

Let $I = [a, b] \subset \mathbb{R}$, $f : I \to \mathbb{R}$, $X : I \to L_2(\Omega)$ and let p be a partition of I and $P = \{p\}$ the set of all partitions of I.

Definition. *The variation of X on I with respect to p is defined as*

$$V_X(p) = \sum_{i=1}^{k} \|X(t_i) - X(t_{i-1})\| \tag{2.47}$$

and the *total variation of X on I* as

$$V_X(I) = \sup_{p \in P} V_X(p) . \tag{2.48}$$

Definition. X is said to be of *bounded variation on I in the strong sense* if $V_X(I)$ is finite.

Example 2.1. Mean-square continuous differentiability on I implies bounded variation in the strong sense on I. For, with p being the (arbitrary) partition of I, Theorems 2.22, 25 give

$$V_X(p) = \sum_{i=1}^{k} \|X(t_i) - X(t_{i-1})\| = \sum_{i=1}^{k} \left\| \int_{t_{i-1}}^{t_i} X'(s)\, ds \right\|$$

$$\leqslant \sum_{i=1}^{k} \int_{t_{i-1}}^{t_i} \|X'(s)\|\, ds \leqslant M(b-a) , \qquad M = \max_{t \in [a,b]} \|X'(t)\| .$$

Example 2.2. We shall show that the Wiener-Lévy process $W : [0, T] \to L_2(\Omega)$ is not of bounded variation on $[0, T]$ in the strong sense. Let the partition p_n of $[0, T]$ have the subdivision points

$$0, \frac{1}{n}T, \ldots, \frac{n-1}{n}T, T .$$

Then, according to (1.79),

$$V_W(p_n) = \sum_{k=1}^{n} \left\| W\left(\frac{k}{n}T\right) - W\left(\frac{k-1}{n}T\right) \right\| = \sum_{k=1}^{n} \sqrt{\frac{1}{n}T}$$

$$= \sqrt{nT} \to \infty \quad \text{as} \quad n \to \infty .$$

Theorem 2.26. If f is continuous and X is of bounded variation on I in the strong sense, then

$$\int_a^b f(t)\, dX(t) \quad \left[\text{hence also} \int_a^b X(t)\, df(t)\right]$$

exists. The integrals have all the properties discussed in Sect. 2.4.1.

Proof. The proof is analogous to that for (a) of Theorem 2.22. □

Theorem 2.27. If f is continuous and X m.s. differentiable with m.s. continuous derivative $X'(t)$ on I, then the integrals given below exist and are identical:

$$\int_a^b f(t)\,dX(t) = \int_a^b f(t)X'(t)\,dt\,. \tag{2.49}$$

Proof. Theorem 2.26 gives the existence of the left-hand integral of (2.49). The right-hand integral also exists as a consequence of Theorem 2.22 since $f(t)X'(t)$ is m.s. continuous on I. To show the equality, let $\{p_n\}_{n\in N}$ be a convergent sequence of partitions of I and let

$$\{S_{f,X}(p_n)\}_{n\in N} \quad \text{and} \quad \{S_{fX'}(p_n)\}_{n\in N}$$

be the corresponding sequences of R–S sums belonging to the left- and right-hand integrals of (2.49), respectively. Then

$$\begin{aligned}
&\left\| \int_a^b f(t)\,dX(t) - \int_a^b f(t)X'(t)\,dt \right\| \\
&\quad \leqslant \left\| \int_a^b f(t)\,dX(t) - S_{f,X}(p_n) \right\| + \|S_{f,X}(p_n) - S_{fX'}(p_n)\| \\
&\quad + \left\| S_{fX'}(p_n) - \int_a^b f(t)X'(t)\,dt \right\| .
\end{aligned}$$

Because of the existence of the integrals, the first and third terms given above can be made arbitrarily small by taking n sufficiently large. We shall show that the same applies to the middle term.

Let p_n be the partition of I. We obtain

$$\begin{aligned}
&\|S_{f,X}(p_n) - S_{fX'}(p_n)\| \\
&\quad = \left\| \sum_{i=1}^k f(t_i')[X(t_i) - X(t_{i-1})] - \sum_{i=1}^k f(t_i')X'(t_i')(t_i - t_{i-1}) \right\| \\
&\quad = \left\| \sum_{i=1}^k f(t_i')\{[X(t_i) - X(t_{i-1})] - X'(t_i')(t_i - t_{i-1})\} \right\| \\
&\quad = \left\| \sum_{i=1}^k f(t_i')\left[\int_{t_{i-1}}^{t_i} X'(t)\,dt - \int_{t_{i-1}}^{t_i} X'(t_i')\,dt \right] \right\| \\
&\quad = \left\| \sum_{i=1}^k f(t_i') \int_{t_{i-1}}^{t_i} [X'(t) - X'(t_i')]\,dt \right\| \\
&\quad \leqslant \sum_{i=1}^k |f(t_i')| \int_{t_{i-1}}^{t_i} \|X'(t) - X'(t_i')\|\,dt\,.
\end{aligned}$$

In the above, the third equality follows because $X'(t_i')$ is a constant element of $L_2(\Omega)$; the inequality at the end is a result of Theorem 2.22.

Since $X'(t)$ is m.s. continuous on I, it is m.s. uniformly continuous on I. Hence, for each $\varepsilon > 0$, there is a $\delta > 0$ such that

$$\|X'(t) - X'(t_i)\| < \varepsilon$$

whenever $|t - t'| < \delta$. Thus, if n is sufficiently large, $\Delta(p_n) < \delta$, and finally

$$\sum_{i=1}^{k} |f(t_i')| \int_{t_{i-1}}^{t_i} \|X'(t) - X'(t_i')\| \, dt \leq M \sum_{i=1}^{k} \int_{t_{i-1}}^{t_i} \varepsilon \, dt$$

$$= M \int_a^b \varepsilon \, dt = M\varepsilon(b - a)$$

where $M = \max_{t \in I} |f(t)|$. This completes the proof. □

2.4.4 A Weak Condition for Existence

Before proceeding, some extensions of the notion of partition disussed at the beginning of Sect. 2.4 are in order. Consider intervals $[a, b]$ and $[c, d]$ of $\mathbb{R}$ and let $R = [a, b] \times [c, d]$ be a rectangle of $\mathbb{R}^2$. If

$$p_s = \{[s_{i-1}, s_i], i = 1, \ldots, m\} \cup \{s_i', i = 1, \ldots, m\}$$

is a partition of $[a, b]$ and

$$p_t = \{[t_{j-1}, t_j], j = 1, \ldots, n\} \cup \{t_j', j = 1, \ldots, n\}$$

a partition of $[c, d]$, then

$$p = p_s \times p_t = \{r_{ij} = [s_{i-1}, s_i] \times [t_{j-1}, t_j], i = 1, \ldots, m; j = 1, \ldots, n\}$$
$$\cup \{(s_i', t_j'), i = 1, \ldots, m; j = 1, \ldots, n\} \tag{2.50}$$

is defined as a *partition of R* (Fig. 2.3). Furthermore, P is used to denote the set of all partitions of R.

A partition $p' = p_s' \times p_t'$ of R is called a *refinement of p* if p_s' and p_t' are refinements of p_s and p_t, respectively. If R is a square $[a, b]^2$, any partition $p_s \times p_t$ may be refined by a partition of the type $p \times p$, where the subdivision points of both p_s and p_t are also subdivision points of p.

The quantity

$$\Delta p = \max(\Delta p_s, \Delta p_t) \tag{2.51}$$

is called the *mesh of p*. A sequence $\{p_n\}_{n \in N}$ of partitions of R is called *convergent* if $\Delta p_n \to 0$ as $n \to \infty$.

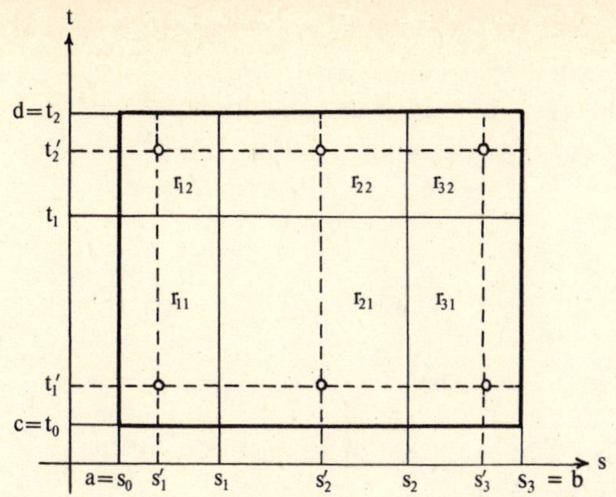

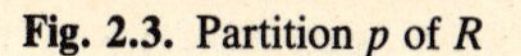
Fig. 2.3. Partition p of R

Let the mapping $G : R \to \mathbb{R}$ be given. If

$$r = [s, s'] \times [t, t'] ,$$

we define

$$\underset{r}{\Delta\Delta} G(x, y) = \underset{[s,s']}{\Delta} \left\{ \underset{[t,t']}{\Delta} G(x, y) \right\} = \underset{[s,s']}{\Delta} \{G(x, t') - G(x, t)\}$$

$$= G(s', t') - G(s', t) - G(s, t') + G(s, t) . \tag{2.52}$$

For any partition p of R, the *variation of G on R* with respect to p is defined as

$$V_G(p) = \sum_{i=1}^{m} \sum_{j=1}^{n} \left| \underset{r_{ij}}{\Delta\Delta} G(x, y) \right| \tag{2.53}$$

and the *total variation of G on R* as

$$V_G(R) = \sup_{p \in P} V_G(p) . \tag{2.54}$$

If $V_G(R)$ is finite, G is said to be of *bounded variation* on R.

If p' is a refinement of p, where p is a partition of R, then

$$V_G(p) \leqslant V_G(p') . \tag{2.55}$$

Where R is the union of two rectangles R_1 and R_2 having only one common side,

$$V_G(R) = V_G(R_1) + V_G(R_2) . \tag{2.56}$$

If R is a square, the total variation of G on R may be computed by using partitions of the type $p \times p$ as

$$V_G(R) = \sup_{p\times p\in P} V_G(p\times p)\,. \tag{2.57}$$

Equation (2.57) follows from the fact that since any partition $p_s \times p_t$ of R may be refined by a partition of type $p \times p$, as shown in Fig. 2.4, (2.55) gives

$$V_G(p \times p) \geqslant V_G(p_s \times p_t)$$

and hence

$$\sup_{p\times p\in P} V_G(p \times p) \geqslant V_G(R)\,.$$

Now, the inverse inequality is also true as the set of all partitions of the type $p \times p$ is a subset of P, so (2.57) thus follows.

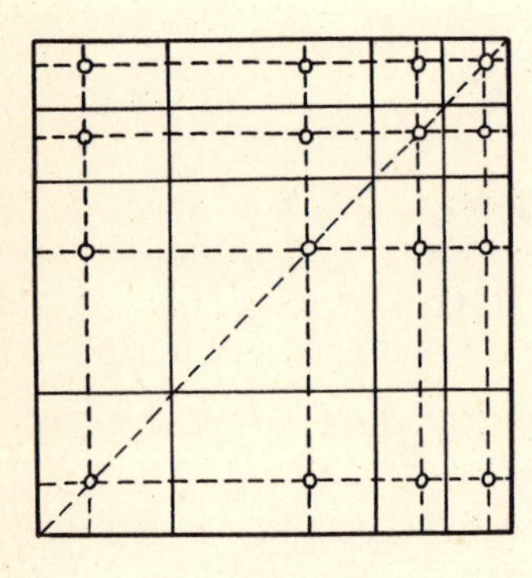

Fig. 2.4. Partition $p \times p$

Let $R = [a, b] \times [c, d]$, $F : R \to \mathbb{R}$, $G : R \to \mathbb{R}$, and let p be the partition of R as defined by (2.50). For this partition we define the *Riemann-Stieltjes* (R–S) *sum*

$$S(p) = \sum_{i=1}^{m} \sum_{j=1}^{n} F(s_i', t_j')\underset{r_{ij}}{\Delta\Delta} G(s, t)\,. \tag{2.58}$$

If for any convergent sequence $\{p_n\}_{n\in\mathbb{N}}$ of partitions of R the corresponding sequence of R–S sums

$$\{S(p_n)\}_{n\in\mathbb{N}}$$

converges, then all these sequences have one and the same limit. This limit is called the *R–S integral* of F on R with respect to G and is denoted by

$$\iint_R F(s, t)\, d_s\, d_t\, G(s, t)\,. \tag{2.59}$$

We remark that this R–S integral definition differs from the usual and more general definition in that the "intermediate" points (s_{ij}', t_{ij}') in the rectangles r_{ij} occurring in the usual R–S sums do not necessarily lie on a grid, as the points (s_i', t_j') do.

Along the lines analogous to those in the proof of Theorem 2.22, we can prove the following result.

Theorem 2.28. If $F : R \to \mathbb{R}$ is continuous and $G : R \to \mathbb{R}$ is of bounded variation on R, then the R–S integral

$$\iint_R F(s, t)\, d_s\, d_t\, G(s, t)$$

exists, and

$$\left| \iint_R F(s, t)\, d_s\, d_t\, G(s, t) \right| \leqslant M V_G(R) , \quad \text{where} \tag{2.60}$$

$$M = \max_{(s,t) \in R} |F(s, t)| .$$

We note that this theorem remains true under the usual and more general R–S integral definition. □

Let us now return to m.s. calculus. Let $I = [a, b] \subset \mathbb{R}$, $X : I \to L_2(\Omega)$ and $C_X(s, t) = \mathrm{E}\{X(s)X(t)\}$, $(s, t) \in I^2$. If $[s, s']$ and $[t, t']$ are intervals of $[a, b]$ and $r = [s, s'] \times [t, t']$, we note that

$$\begin{aligned} \underset{r}{\Delta\Delta} C_X(s, t) &= \mathrm{E}\{X(s')X(t')\} - \mathrm{E}\{X(s')X(t)\} - \mathrm{E}\{X(s)X(t')\} \\ &\quad + \mathrm{E}\{X(s)X(t)\} \\ &= \mathrm{E}\{[X(s') - X(s)][X(t') - X(t)]\} . \end{aligned} \tag{2.61}$$

Definition. If the correlation function C_X is of bounded variation on I^2, then X is said to be of *bounded variation on I in the weak sense.*

The condition under which X is of bounded variation in the weak sense is weaker than that for X to be of bounded variation in the strong sense. This follows from

$$V_{C_X}(I^2) \leqslant [V_X(I)]^2 . \tag{2.62}$$

To show this inequality, let p be the partition of I defined by (2.16). Then $p \times p$ is a partition of I^2, and (2.61) together with the Schwartz inequality leads to

$$\begin{aligned} V_{C_X}(p \times p) &= \sum_{i=1}^{k} \sum_{j=1}^{k} |\mathrm{E}\{[X(s_i) - X(s_{i-1})][X(s_j) - X(s_{j-1})]\}| \\ &\leqslant \left\{ \sum_{i=1}^{k} \|X(s_i) - X(s_{i-1})\| \right\}^2 = [V_X(p)]^2 , \end{aligned}$$

giving (2.62) in view of (2.57).

Example 2.3. We have shown in Example 2.2 that the Wiener-Lévy process $W: [0, T] \to L_2(\Omega)$ is not of bounded variation on $[0, T]$ in the strong sense. We shall show here that it is of bounded variation in the weak sense on $[0, T]$ with

$$V_{C_W}([0, T]^2) = T . \tag{2.63}$$

To prove our assertion, we see from (2.57) that we may compute $V_{C_W}([0, T]^2)$ by using partitions of $[0, T]^2$ of the type $p \times p$. Suppose p has subdivision points $t_o, t_1, \ldots, t_n$ satisfying

$$0 = t_o < t_1 \ldots < t_n = T .$$

Then the rectangles of $p \times p$ are

$$r_{ij} = [t_{i-1}, t_i] \times [t_{j-1}, t_j] , \quad i, j = 1, \ldots, n .$$

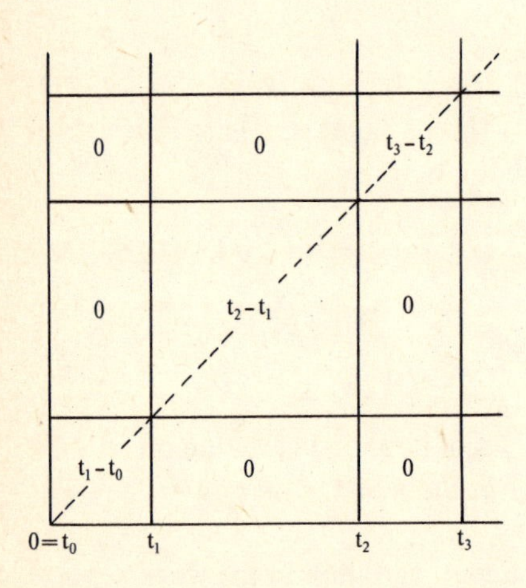

Fig. 2.5. Values of $\underset{r_{ij}}{\Delta\Delta} C_W(s, t)$

With the properties of W discussed in Sect. 1.4.4, (2.52) leads to (Fig. 2.5)

$$\begin{aligned}\underset{r_{ij}}{\Delta\Delta} C_W(s, t) &= E\{[W(t_i) - W(t_{i-1})][W(t_j) - W(t_{j-1})]\} \\ &= \begin{cases} E\{W(t_i) - W(t_{i-1})\}^2 = t_i - t_{i-1} , & i = j \\ 0 & i \neq j \end{cases} .\end{aligned}$$

Hence,

$$\begin{aligned}V_{C_W}(p \times p) &= \sum_{i=1}^{n} \sum_{j=1}^{n} \left| \underset{r_{ij}}{\Delta\Delta} C_W(s, t) \right| = \sum_{i=1}^{n} \left| \underset{r_{ii}}{\Delta\Delta} C_W(s, t) \right| \\ &= \sum_{i=1}^{n} (t_i - t_{i-1}) = T - 0 = T .\end{aligned}$$

Since this result is independent of the partition $p \times p$, (2.63) follows.

Theorem 2.29. If $f: I \to \mathbb{R}$ is continuous and $X: I \to L_2(\Omega)$ of bounded variation on I, in the weak sense, then, as $I = [a, b]$,

$$\int_a^b f(t)\, dX(t) \quad \left[\text{hence also } \int_a^b X(t)\, df(t)\right] \quad \text{and}$$

$$\iint_{I^2} f(s) f(t)\, d_s d_t C_X(s, t)$$

exist, and

$$\left\| \int_a^b f(t)\, dX(t) \right\|^2 = \iint_{I^2} f(s) f(t)\, d_s d_t C_X(s, t)$$
$$\leqslant M^2 V_{C_X}(I^2)\,, \quad \text{where} \tag{2.64}$$

$$M = \max_{t \in I} |f(t)| \quad \text{and} \quad C_X(s, t) = \mathrm{E}\{X(s) X(t)\}\,, \quad (s, t) \in I^2\,.$$

These m.s. integrals have all the properties discussed in Sect. 2.4.1.

Proof. Let $\{p_n\}_{n \in \mathbb{N}}$ be a convergent sequence of partitions of I. For the existence of $\int_a^b f(t)\, dX(t)$, it is sufficient to show that the sequence of R–S sums

$$\{S_{f,X}(p_n)\}_{n \in \mathbb{N}}$$

is a Cauchy sequence in $L_2(\Omega)$, or, equivalently, that

$$\mathrm{E}\{S_{f,X}(p_m)\, S_{f,X}(p_n)\}$$

converges as $m, n \to \infty$ [see (C) of Sect. 2.1].

Let $s_o, s_1, \ldots, s_q$ and $t_o, t_1, \ldots, t_r$ be the subdivision points of p_m and p_n, respectively, such that

$$a = s_o < s_1 < \ldots < s_q = b \quad \text{and} \quad a = t_o < t_1 < \ldots < t_r = b$$

and s_i' and t_j', $i = 1, \ldots, q$, $j = 1, \ldots, r$, be the intermediate points of p_m and p_n, respectively, such that

$$s_i' \in [s_{i-1}, s_i] \quad \text{and} \quad t_j' \in [t_{j-1}, t_j]\,.$$

Then,

$$\mathrm{E}\{S_{f,X}(p_m)\, S_{f,X}(p_n)\} = \mathrm{E}\left\{\left[\sum_{i=1}^{q} f(s_i')[X(s_i) - X(s_{i-1})]\right] \times \left[\sum_{j=1}^{r} f(t_j')[X(t_j) - X(t_{j-1})]\right]\right\}$$
$$= \sum_{i=1}^{q} \sum_{j=1}^{r} f(s_i') f(t_j')\, \underset{r_{ij}}{\Delta\Delta}\, C_X(s, t)\,, \quad \text{where}$$

$$r_{ij} = [s_{i-1}, s_i] \times [t_{j-1}, t_j]\,.$$

The sum on the right-hand side of the equality given above is a R–S sum of the integral

$$\iint_{I^2} f(s) f(t)\, d_s d_t C_X(s, t)$$

whose existence is assured by the conditions imposed on f and C_X. Hence, as $m, n \to \infty$,

$$\mathrm{E}\{S_{f,X}(p_m) S_{f,X}(p_n)\} \to \iint_{I^2} f(s) f(t)\, d_s d_t C_X(s, t) . \tag{2.65}$$

Equation (2.65) thus implies that as $n \to \infty$, $S_{f,X}(p_n)$ converges in m.s. to

$$\int_a^b f(t)\, dX(t) .$$

Then, following (B) of Sect. 2.1,

$$\mathrm{E}\{S_{f,X}(p_m) S_{f,X}(p_n)\} \to \mathrm{E}\left\{\int_a^b f(t)\, dX(t)\right\}^2 \tag{2.66}$$

as $m, n \to \infty$. The theorem now follows from (2.65, 66) and from Theorem 2.28. □

Example 2.4. As a direct consequence of Example 2.3 and Theorem 2.29, we have the following result for the Wiener-Lévy process.

If $f : [0, T] \to \mathbb{R}$ is continuous on $[0, T]$, the integral

$$\int_0^T f(t)\, dW(t)$$

exists, where $W : [0, T] \to L_2(\Omega)$ is the (standard) Wiener-Lévy process.

Theorem 2.30. If $f : [a, b] \to \mathbb{R}$ and $g : [c, d] \to \mathbb{R}$ are continuous on their respective domains and if $X : [a, b] \to L_2(\Omega)$ and $Y : [c, d] \to L_2(\Omega)$ are of bounded variation in the weak sense on their respective domains, then, with $R = [a, b] \times [c, d]$,

$$\iint_R f(s) g(t)\, d_s d_t \mathrm{E}\{X(s) Y(t)\}$$

exists in the sense defined in this section,

$$\int_a^b f(s)\, dX(s) \quad \text{and} \quad \int_c^d g(t)\, dY(t)$$

exist, and

$$\mathrm{E}\left\{\int_a^b f(s)\, dX(s) \int_c^d g(t)\, dY(t)\right\} = \iint_R f(s) g(t)\, d_s d_t \mathrm{E}\{X(s) Y(t)\} . \tag{2.67}$$

Proof. Let $\{p_k\}_{k\in N}$ be a convergent sequence of partitions of R and let $p_k = p_{k,s} \times p_{k,t}$ be the partition $p_s \times p_t$ defined in (2.50). Then, because of Theorem 2.29,

$$S_{f,X}(p_{k,s}) = \sum_{i=1}^{m} f(s_i')[X(s_i) - X(s_{i-1})] \to \int_a^b f(s)\, dX(s) \text{ in m.s. as } k \to \infty$$

$$S_{g,Y}(p_{k,t}) = \sum_{j=1}^{n} g(t_j')[Y(t_j) - Y(t_{j-1})] \to \int_c^d g(t)\, dY(t) \text{ in m.s. as } k \to \infty\,.$$

Now

$$\mathrm{E}\{S_{f,X}(p_{k,s})\, S_{g,Y}(p_{k,t})\}$$
$$= \sum_{i=1}^{m} \sum_{j=1}^{n} f(s_i')\, g(t_j')\, \mathrm{E}\{[X(s_i) - X(s_{i-1})][Y(t_j) - Y(t_{j-1})]\}\,. \tag{2.68}$$

It follows from (B) of Sect. 2.1 that the left-hand side of (2.68) converges to

$$\mathrm{E}\left\{\int_a^b f(s)\, dX(s) \int_c^d g(t)\, dY(t)\right\}$$

as $k \to \infty$.

Hence, for each convergent sequence of partitions of R, the right-hand side of (2.68) also converges. Following the discussion on ordinary multiple Riemann-Stieltjes integrals in this section, this limit is

$$\iint_R f(s)\, g(t)\, d_s d_t \mathrm{E}\{X(s) X(t)\}\,.$$

The proof is thus complete. □

Theorem 2.31. Let $I = [a, b] \subset R$. If $f: I \to R$ and $g: I \to R$ are continuous on I and if $X: I \to L_2(\Omega)$ is of bounded variation on I in the weak sense, then

a) $Y(t) = \int_a^t g(s)\, dX(s)$ is defined at each $t \in I$;

b) $Y: I \to L_2(\Omega)$ is of bounded variation on I in the weak sense;

c) $\int_a^b f(t)\, dY(t)$ exists and

$$\int_a^b f(t)\, dY(t) = \int_a^b f(t)\, g(t)\, dX(t)\,. \tag{2.69}$$

Proof. Part (a) has been shown in Theorem 2.29. Concerning (b) (see also Theorem 2.29), set

$$C_X(s, t) = \mathrm{E}\{X(s) X(t)\}\,, \quad (s, t) \in I^2 \quad \text{and}$$

$$C_Y(s, t) = \mathrm{E}\{Y(s) Y(t)\} = \iint_{[a,s]\times[a,t]} g(u)\, g(v)\, d_u d_v C_X(u, v),$$
$$(s, t) \in I^2\,.$$

We need to show that C_Y is of bounded variation on I^2. Let $r = [x, y] \times [w, z]$ be one of the rectangles of a partition $p = \{r\}$ of I^2. Then, according to (2.61) and Theorem 2.29,

$$\begin{aligned} \left|\underset{r}{\Delta\Delta} C_Y(s, t)\right| &= |\mathrm{E}\{[Y(y) - Y(x)][Y(z) - Y(w)]\}| \\ &= \left|\mathrm{E}\left\{\int_x^y g(s)\, dX(s) \int_w^z g(t)\, dX(t)\right\}\right| \\ &= \left|\iint_r g(s) g(t)\, d_s d_t C_X(s, t)\right| \\ &\leqslant m^2 V_{C_X}(r)\,, \quad \text{where} \end{aligned} \tag{2.70}$$

$$m = \max_{s \in [a, b]} |g(s)|\,.$$

Equations (2.56, 70) then lead to

$$V_{C_Y}(p) = \sum_{r \in p} \left|\underset{r}{\Delta\Delta} C_Y(s, t)\right| \leqslant \sum_{r \in p} m^2 V_{C_X}(r) = m^2 V_{C_X}(p)$$

and hence also

$$V_{C_Y}(I^2) \leqslant m^2 V_{C_X}(I^2) < \infty\,.$$

To prove (c), we first remark that the integrals in (2.69) exist as a result of (b) and Theorem 2.29. Let $\{p_n\}_{n \in N}$ be any convergent sequence of partitions of I, let p_n be the partition with subdivision points $t_o, t_1, \ldots, t_k$ such that

$$a = t_o < t_1 < \ldots < t_k = b$$

and with intermediate points t_i', $t_i' \in [t_{i-1}, t_i]$, $i = 1, \ldots, k$, and let $r_{ij} = [t_{i-1}, t_i] \times [t_{j-1}, t_j]$. Denoting the R–S sums with respect to p_n corresponding to the integrals in (2.69) by $S_{f,Y}(p_n)$ and $S_{fg,X}(p_n)$, respectively, we have, with $\varepsilon > 0$ and n sufficiently large,

$$\begin{aligned} &\|S_{f,Y}(p_n) - S_{fg,X}(p_n)\|^2 \\ &= \left\|\sum_{i=1}^k f(t_i')\left\{\int_{t_{i-1}}^{t_i} g(s)\, dX(s) - g(t_i')[X(t_i) - X(t_{i-1})]\right\}\right\|^2 \\ &= \mathrm{E}\left\{\sum_{i=1}^k f(t_i')\left[\int_{t_{i-1}}^{t_i} [g(s) - g(t_i')]\, dX(s)\right]\right\}^2 \\ &= \left|\sum_{i=1}^k \sum_{j=1}^k f(t_i') f(t_j') \iint_{r_{ij}} [g(s) - g(t_i')][g(t) - g(t_j')]\, d_s d_t \mathrm{E}\{X(s) X(t)\}\right| \\ &\leqslant M^2 \sum_{i=1}^k \sum_{j=1}^k \varepsilon^2 V_{C_X}(r_{ij}) = M^2 \varepsilon^2 V_{C_X}(I^2)\,, \quad \text{where} \end{aligned}$$

$$M = \max_{t \in I} |f(t)|\,.$$

In the above, we have used the fact that

$$|g(t) - g(t_i')| < \varepsilon \quad \text{as} \quad t \in [t_{i-1}, t_i], \quad i = 1, \ldots, k .$$

This is true (because of the uniform continuity of g on I) if n is sufficiently large. We thus see that

$$\|S_{f,Y}(p_n) - S_{fg,X}(p_n)\|$$

is arbitrarily small if n is sufficiently large, completing the proof. □

2.4.5 Supplementary Exercises

Exercise 2.3. Let $X : [0, 1] \to L_2(\Omega)$ and $f : [0, 1] \to \mathbb{R}$ be defined as

$$X(t) = \begin{cases} Z, & \text{if} \quad t = \frac{1}{2} \\ 0, & \text{if} \quad t \in [0, 1] \backslash \{\frac{1}{2}\} \end{cases} \tag{2.71}$$

$$f(t) = \begin{cases} 0, & \text{if} \quad t \in [0, \frac{1}{2}] \\ 1, & \text{if} \quad t \in (\frac{1}{2}, 1] \end{cases}, \tag{2.72}$$

where $Z \in L_2(\Omega)$ with $\mathrm{E}Z^2 = 1$. Show that

$$\int_0^1 f(t)\, dX(t) \quad \text{and} \quad \int_0^1 X(t)\, df(t)$$

do not exist.

Exercise 2.4. Let $X : [0, 1] \to L_2(\Omega)$ be an arbitrary second-order process and let $f : [0, 1] \to \mathbb{R}$ be a constant function: $\forall t \in [0, t], f(t) = c \in \mathbb{R}$. Show that both

$$\int_0^1 f(t)\, dX(t) \quad \text{and} \quad \int_0^1 X(t)\, df(t) \tag{2.73}$$

exist and give their respective values.

Exercise 2.5. Verify Theorem 2.17 with the integrals given in (2.73).

Exercise 2.6. Let $f : [0, 2] \to \mathbb{R}$ be defined as follows:

$$f(t) = \begin{cases} 0, & \text{if} \quad t \in [0, 1) \\ 1, & \text{if} \quad t \in [1, 2] \end{cases} . \tag{2.74}$$

Let C be a nonzero element of $L_2(\Omega)$ and let $X : [0, 2] \to L_2(\Omega)$ be defined as

$$X(t) = \begin{cases} 0, & \text{if } t \in [0, 1] \\ C, & \text{if } t \in (1, 2] \end{cases} . \tag{2.75}$$

Show that

$$\int_0^1 f(t)\, dX(t) \quad \text{and} \quad \int_1^2 f(t)\, dX(t)$$

exist but that

$$\int_0^2 f(t)\, dX(t)$$

does not.

Exercise 2.7. Let a, b, and c be real numbers such that $a < c < b$ and let $f : [a, b] \to \mathbb{R}$ and $X : [a, b] \to L_2(\Omega)$ be such that

$$\int_a^b f(t)\, dX(t)$$

exists. Show that

a) $\int_a^c f(t)\, dX(t)$ and $\int_c^b f(t)\, dX(t)$ exist;

b) $\int_a^b f(t)\, dX(t) = \int_a^c f(t)\, dX(t) + \int_c^b f(t)\, dX(t)$.

Exercise 2.8. Let $I = [a, b] \subset \mathbb{R}$ and $X : I \to L_2(\Omega)$. Show that $EX : I \to \mathbb{R}$ is of bounded variation on I if X is of bounded variation on I in the strong sense.

Exercise 2.9. Let R be the rectangle $[a, b] \times [c, d]$ in $\mathbb{R}^2$. Let r_i, $i = 1, \ldots, n$, be rectangles in R with sides parallel to those of R such that

$$R = \bigcup_{i=1}^{n} r_i$$

and such that r_i and r_j $(i \neq j)$ have at most one common side (Fig. 2.6). Let

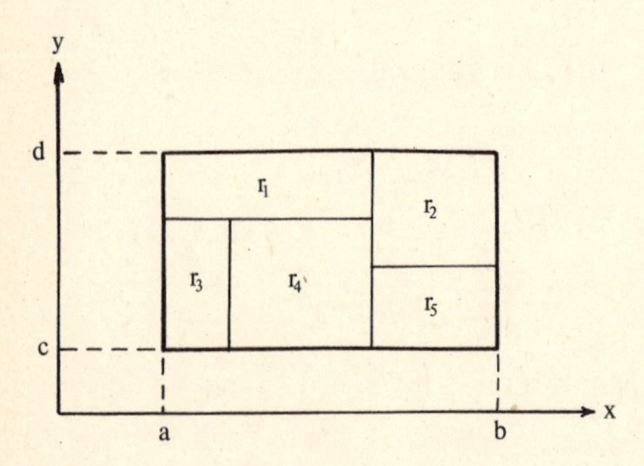

Fig. 2.6. Rectangles R and r_i, $i = 1, \ldots, n$

$q = \{r_1, \ldots, r_n\}$ and let $Q = \{q\}$ be the set of sets of type q. If $F : R \to \mathbb{R}$, show that the total variation of F on R is equal to

$$\sup_{q \in Q} W_F(q) , \quad \text{where} \quad W_F(q) = \sum_{i=1}^{n} \left| \underset{r_i}{\Delta\Delta} F(x, y) \right| .$$

Exercise 2.10. Let $[a, b] \subset \mathbb{R}$, $[c, d] \subset \mathbb{R}$, $f : [a, b] \to \mathbb{R}$, $g : [c, d] \to \mathbb{R}$, $V_f([a, b]) < \infty$, $V_g([c, d]) < \infty$, and let

$$H(s, t) = f(s) g(t), \ (s, t) \in R = [a, b] \times [c, d] .$$

Show that H is of bounded variation on R with total variation

$$V_H(R) = V_f([a, b]) \, V_g([c, d]) .$$

Exercise 2.11. Let $I : [a, b] \subset \mathbb{R}$ and $X : I \to L_2(\Omega)$. Show that $\mathrm{E}X : I \to \mathbb{R}$ is of bounded variation on I if X is of bounded variation on I in the weak sense.

2.5 Mean-Square Calculus of Random N Vectors

In this section, we consider extensions of some of the previous results to the case of random-vector processes.

Let $I = [a, b] \subset \mathbb{R}$, $X_i : I \to L_2(\Omega)$, $i = 1, \ldots, N$, and let $L_2^N(\Omega)$ be the Cartesian product of N identical spaces $L_2(\Omega)$. Let us define $\boldsymbol{X} : I \to L_2^N(\Omega)$ as

$$\boldsymbol{X}(t) = \begin{bmatrix} X_1(t) \\ \vdots \\ X_N(t) \end{bmatrix} , \quad t \in I . \tag{2.76}$$

If X_i is m.s. differentiable at $t \in I$ (or on I), $i = 1, \ldots, N$, then $\boldsymbol{X}$ is said to be m.s. differentiable at t (or on I), and its m.s. derivative is defined by

$$\boldsymbol{X}'(t) = \begin{bmatrix} X_1'(t) \\ \vdots \\ X_N'(t) \end{bmatrix} , \quad t \in I . \tag{2.77}$$

As a consequence of the results in Sect. 2.3, we have

$$\mathrm{E}\boldsymbol{X}'(t) = \frac{d}{dt} \mathrm{E}\boldsymbol{X}(t) \tag{2.78}$$

and, if $J = [c, d]$ and $Y : J \to L_2^N(\Omega)$ is also m.s. differentiable, the following also hold:

$$\begin{aligned} E\{X'(s)Y^T(t)\} &= \frac{\partial}{\partial s}E\{X(s)Y^T(t)\}\ ; \\ E\{X(s)Y^{T\prime}(t)\} &= \frac{\partial}{\partial t}E\{X(s)Y^T(t)\}\ ; \\ E\{X'(s)Y^{T\prime}(t)\} &= \frac{\partial^2}{\partial s\,\partial t}E\{X(s)Y^T(t)\}\ . \end{aligned} \tag{2.79}$$

Let $F(t) = [f_{ij}(t)]$, $t \in I$, be an $M \times N$ matrix. If

$$\int_a^b f_{ij}(t)\,X_j(t)\,dt$$

exists for all i and j, we define

$$\int_a^b F(t)X(t)\,dt = \begin{bmatrix} \int_a^b f_{11}(t)X_1(t)\,dt & + \ldots + & \int_a^b f_{1N}(t)X_N(t)\,dt \\ & \vdots & \\ \int_a^b f_{M1}(t)X_1(t)\,dt & + \ldots + & \int_a^b f_{MN}(t)X_N(t)\,dt \end{bmatrix}. \tag{2.80}$$

In particular,

$$\int_a^b X(t)\,dt = \begin{bmatrix} \int_a^b X_1(t)\,dt \\ \vdots \\ \int_a^b X_N(t)\,dt \end{bmatrix}. \tag{2.81}$$

If

$$\int_a^b X_j(t)\,df_{ij}(t) \tag{2.82}$$

exists for all i and j, we define

$$\int_a^b [dF(t)]X(t) = \begin{bmatrix} \int_a^b X_1(t)\,\mathrm{df}_{11}(t) & + \ldots + & \int_a^b X_N(t)\,df_{1N}(t) \\ & \vdots & \\ \int_a^b X_1(t)\,df_{M1}(t) & + \ldots + & \int_a^b X_N(t)\,df_{MN}(t) \end{bmatrix}. \tag{2.83}$$

Similarly, if

$$\int_a^b f_{ij}(t)\,dX_j(t) \tag{2.84}$$

exists for all i and j, we define

$$\int_a^b F(t)\,dX(t) = \begin{bmatrix} \int_a^b f_{11}(t)\,dX_1(t) + \ldots + \int_a^b f_{1N}(t)\,dX_N(t) \\ \vdots \\ \int_a^b f_{M1}(t)\,dX_1(t) + \ldots + \int_a^b f_{MN}(t)\,dX_N(t) \end{bmatrix} . \tag{2.85}$$

2.5.1 Conditions for Existence

A sufficient condition for the existence of (2.83, 85) is that X_i is m.s. continuous on I and f_{ij} is of bounded variation on I, $i = 1, \ldots, N$ and $j = 1, \ldots, M$, or that f_{ij} is continuous on I and X_i is of bounded variation on I in the weak (or strong) sense, $i = 1, \ldots, N$ and $j = 1, \ldots, M$. When they exist, the expectations of (2.83, 85) exist and are, respectively,

$$\mathrm{E}\left\{\int_a^b [dF(t)]\,X(t)\right\} = \int_a^b [dF(t)]\,\mathrm{E}X(t) \tag{2.86}$$

$$\mathrm{E}\left\{\int_a^b F(t)\,dX(t)\right\} = \int_a^b F(t)\,d\mathrm{E}X(t) . \tag{2.87}$$

Let $J = [c, d] \subset \mathbb{R}$, $Y: J \to L_2^M(\Omega)$, and let G be a $N \times M$ matrix with continuous elements $g_{km}: J \to \mathbb{R}$, $k = 1, \ldots, N$ and $m = 1, \ldots, M$. Suppose that X_i and Y_k are of bounded variation on I and J, respectively, in the weak sense and that the elements f_{ij} of F are continuous on I. Then, as a consequence of Theorem 2.30,

$$\mathrm{E}\left\{\int_a^b F(s)\,dX(s)\left[\int_c^d G(t)\,dY(t)\right]^T\right\}$$
$$= \iint_R F(s)\,[d_s d_t \mathrm{E}\{X(s)\,Y^T(t)\}]\,G^T(t) , \tag{2.88}$$

where $R = I \times J$. It follows from the fact that the ith element of

$$\int_a^b F(s)\,dX(s) \quad \text{is} \quad \sum_{j=1}^{N} \int_a^b f_{ij}(s)\,dX_j(s)$$

and the kth element of

$$\int_c^d G(t)\,dY(t)$$

is equal to

$$\sum_{m=1}^{M} \int_c^d g_{km}(t)\,dY_m(t) .$$

Thus, the ikth element of the $M \times N$ matrix in (2.88) takes the form

$$\sum_{j=1}^{N} \sum_{m=1}^{M} \iint_R f_{ij}(s)\,g_{km}(t)\,d_s d_t \mathrm{E}\{X_j(s)\,Y_m(t)\} .$$

2.6 The Wiener-Lévy Process

Because of the central place it occupies in our discourse, the Wiener-Lévy process is now considered in more detail in the context of m.s. calculus. Let $W : [0, T] \to L_2(\Omega)$ be the standard Wiener-Lévy process. We recall that some of its m.s. properties have been discussed in Sect. 2.3, 4:

a) it is nowhere m.s. differentiable;
b) it is not of bounded variation on $[0, T]$ in the strong sense;
c) it is of bounded variation on $[0, T]$ in the weak sense;
d) let $f : [0, T] \to \mathbb{R}$. A sufficient condition for existence of

$$\int_0^T f(t)\, dW(t)$$

is the continuity of f on $[0, T]$.

Theorem 2.32. If f and $g : [0, T] \to \mathbb{R}$ are continuous on $[0, T]$ and if $[a, b] \subset [0, T]$, then

$$\int_a^b f(t)\, dW(t) \quad \text{and} \quad \int_a^b g(t)\, dW(t)$$

exist and

$$\mathrm{E}\left\{\int_a^b f(u)\, dW(u) \int_a^b g(v)\, dW(v)\right\} = \int_a^b f(t)\, g(t)\, dt\,. \tag{2.89}$$

In particular,

$$\left\|\int_a^b f(t)\, dW(t)\right\|^2 = \mathrm{E}\left\{\int_a^b f(t)\, dW(t)\right\}^2 = \int_a^b f^2(t)\, dt\,. \tag{2.90}$$

Proof. Since all integrals given above exist (Theorem 2.29), they may all be constructed as m.s. limits of R–S sums defined on the same convergent sequence $\{p_n\}_{n\in\mathbb{N}}$ of partitions of $[a, b]$. According to (B) of Sect. 2.1, we thus have

$$\begin{aligned}\mathrm{E}\left\{\int_a^b f(u)\, dW(u) \int_a^b g(v)\, dW(v)\right\} &= \mathrm{E}\left\{\lim_{m\to\infty} S_{f,W}(p_m) \lim_{n\to\infty} S_{g,W}(p_n)\right\} \\ &= \lim_{n\to\infty} \mathrm{E}\left\{S_{f,W}(p_n)\, S_{g,W}(p_n)\right\}\,,\end{aligned}$$

and, with p_n as described in Sect. 2.4,

$$\begin{aligned}\mathrm{E}\left\{S_{f,W}(p_n)\, S_{g,W}(p_n)\right\} &= \sum_{i=1}^{k} \sum_{j=1}^{k} f(t_i')\, g(t_j') \\ &\quad \times \mathrm{E}\{[W(t_i) - W(t_{i-1})][W(t_j) - W(t_{j-1})]\} \\ &= \sum_{i=1}^{k} f(t_i')\, g(t_i')(t_i - t_{i-1})\end{aligned}$$

as a consequence of the orthogonality of increments of W and its other properties, discussed in Sect. 1.4.4.

Since the right-hand side given above is a Riemann sum belonging to

$$\int_a^b f(t)\, g(t)\, dt\,,$$

(2.89) then follows as $n \to \infty$. □

Theorem 2.33. If f and $g : [0, T] \to \mathbb{R}$ are continuous on $[0, T]$ and if s and $t\,(s \leqslant t)$ are in $[0, T]$, then

$$\mathrm{E}\left\{\int_0^s f(u)\, dW(u) \int_0^t g(v)\, dW(v)\right\} = \int_0^s f(u)\, g(u)\, du \tag{2.91}$$

and, in particular,

$$\mathrm{E}\left\{\int_0^s f(u)\, dW(u) \int_0^t f(v)\, dW(v)\right\} = \int_0^s f^2(u)\, du\,. \tag{2.92}$$

Proof. We have

$$\begin{aligned}
&\mathrm{E}\left\{\int_0^s f(u)\, dW(u) \int_0^t g(v)\, dW(v)\right\} \\
&\quad = \mathrm{E}\left\{\int_0^s f(u)\, dW(u)\left[\int_0^s g(v)\, dW(v) + \int_s^t g(v)\, dW(v)\right]\right\} \\
&\quad = \mathrm{E}\left\{\int_0^s f(u)\, dW(u) \int_0^s g(v)\, dW(v)\right] + \mathrm{E}\left\{\int_0^s f(u)\, dW(u) \int_s^t g(v)\, dW(v)\right\}.
\end{aligned}$$

According to the preceding theorem, the first term on the right-hand side is equal to

$$\int_0^s f(u)\, g(u)\, du\,.$$

As to the second term, the R–S sums belonging to

$$\int_0^s f(u)\, dW(u) \quad \text{and} \quad \int_s^t g(v)\, dW(v)$$

are linear combinations of increments of W on $[0, s]$ and $[s, t]$, respectively. Since $[0, s)$ and $[s, t)$ are disjoint, these R–S sums are orthogonal, leading to

$$\mathrm{E}\left\{\int_0^s f(u)\, dW(u) \int_s^t g(v)\, dW(v)\right\} = 0\,,$$

by virtue of (B) of Sect.2.1. □

2.6.1 The General Wiener-Lévy N Vector

Let $\boldsymbol{W}^o : [0, T] \to L_2^N(\Omega)$ be the standard Wiener-Lévy N vector introduced in Sect. 1.4.4. There we have shown

$$\mathrm{E}\boldsymbol{W}^o(t) = 0 \quad \text{and} \quad \mathrm{E}\{\boldsymbol{W}^o(s)\,\boldsymbol{W}^{oT}(t)\} = sI \tag{2.93}$$

if $0 \leqslant s \leqslant t \leqslant T$, where I is the $N \times N$ identity matrix.

Before defining the general Wiener-Lévy N vector, let us first indicate how it may be constructed.

Let $Q(t)$, $t \in [0, T]$, be an $N \times N$ matrix whose elements $q_{ij}(t)$ are real-valued and continuous on $[0, T]$. An N vector $\boldsymbol{W} : [0, T] \to L_2^N(\Omega)$ can be defined as

$$\boldsymbol{W}(t) = \int_0^t Q(s)\,d\boldsymbol{W}^o(s)\,, \qquad t \in [0, T]\,, \tag{2.94}$$

i.e.,

$$\boldsymbol{W}(t) = \begin{bmatrix} W_1(t) \\ \vdots \\ W_N(t) \end{bmatrix} = \begin{bmatrix} \int_0^t q_{11}(s)\,dW_1^o(s) + \ldots + \int_0^t q_{1N}(s)\,dW_N^o(s) \\ \vdots \\ \int_0^t q_{N1}(s)\,dW_1^o(s) + \ldots + \int_0^t q_{NN}(s)\,dW_N^o(s) \end{bmatrix}. \tag{2.95}$$

The first part of (2.93) clearly gives

$$\mathrm{E}\boldsymbol{W}(t) = \boldsymbol{0} \in \mathbb{R}^N\,, \qquad t \in [0, T]\,. \tag{2.96}$$

Furthermore,

$$\mathrm{E}\{\boldsymbol{W}(s)\,\boldsymbol{W}^T(t)\} = \int_0^s Q(u)\,Q^T(u)\,du\,, \tag{2.97}$$

if $0 \leqslant s \leqslant t \leqslant T$. This can be verified by observing that following Theorem 2.33 the ijth element of (2.97) is

$$\begin{aligned} \mathrm{E}\{W_i(s)\,W_j(t)\} &= \mathrm{E}\left\{\sum_{k=1}^N \int_0^s q_{ik}(u)\,dW_k^o(u) \sum_{m=1}^N \int_0^t q_{jm}(v)\,dW_m^o(v)\right\} \\ &= \sum_{k=1}^N \sum_{m=1}^N \mathrm{E}\left\{\int_0^s q_{ik}(u)\,dW_k^o(u) \int_0^t q_{jm}(v)\,dW_m^o(v)\right\} \\ &= \sum_{k=1}^N \int_0^s q_{ik}(u)\,q_{jk}(u)\,du\,. \end{aligned}$$ □

Setting

$$B(u) = Q(u)\,Q^T(u) \tag{2.98}$$

(2.97) becomes

$$\mathrm{E}\{\boldsymbol{W}(s)\boldsymbol{W}^T(t)\} = \int_0^s B(u)\,du \quad \text{if} \quad 0 \leqslant s \leqslant t \leqslant T\,. \tag{2.99}$$

Then the $N \times N$ matrix $B(u)$ has the following properties:

a) its elements $b_{ij}(u)$ are continuous real functions of u on $[0, T]$;
b) $B(u)$ is symmetric at each $u \in [0, T]$;
c) $B(u) \geqslant 0$ at each $u \in [0, T]$, i.e.,

$$\forall \boldsymbol{x} = \begin{bmatrix} x_1 \\ \vdots \\ x_N \end{bmatrix} \in \mathbb{R}^N, \quad \boldsymbol{x}^T B(u)\boldsymbol{x} \geqslant 0\,, \tag{2.100}$$

since

$$\boldsymbol{x}^T B(u)\boldsymbol{x} = [Q(u)\boldsymbol{x}]^T[Q(u)\boldsymbol{x}] = \sum_{i=1}^N \left[\sum_{j=1}^N q_{ij}(u)x_j\right]^2 \geqslant 0\,.$$

Conversely, given an N-vector process $\boldsymbol{W} : [0, T] \to L_2^N(\Omega)$ with

$$\mathrm{E}\boldsymbol{W}(t) = 0 \in \mathbb{R}^N \quad \text{and} \quad \mathrm{E}\{\boldsymbol{W}(s)\boldsymbol{W}^T(t)\} = \int_0^s B(u)\,du\,,$$

if $0 \leqslant s \leqslant t \leqslant T$, where $B(u)$ has the three properties given above, then, at each $u \in [0, T]$, the eigenvalues $\lambda_1(u), \ldots, \lambda_N(u)$ of $B(u)$ are real-valued and nonnegative. Furthermore, there is an $N \times N$ orthogonal matrix $R(u)$ such that

$$B(u) = R(u)\,\mathrm{diag}\,(\lambda_1(u), \ldots, \lambda_N(u))\,R^T(u)\,. \tag{2.101}$$

If we define

$$Q(u) = R(u)\,\mathrm{diag}\,(\sqrt{\lambda_1(u)}, \ldots, \sqrt{\lambda_N(u)})\,R^T(u)\,, \tag{2.102}$$

then (2.101) leads to

$$B(u) = Q(u)\,Q^T(u)\,,$$

where $Q(u)$ is a real-valued (and symmetric) $N \times N$ matrix at each $u \in [0, T]$. Putting

$$\boldsymbol{W}(t) = \int_0^t Q(u)\,d\boldsymbol{W}^o(u)$$

we have reconstructed a representation of $\boldsymbol{W}(t)$ as described in (2.94) except for the continuity of the elements of $Q(u)$ on $[0, T]$, which is a sufficient condition for existence of the representation. While it can be shown that there are continuous matrices $Q(u)$, $u \in [0, T]$, as meant here, we shall omit the tedious argument since it is not needed in the development that follows.

Definition. $W : [0, T] \to L_2^N(\Omega)$ is *the general Wiener-Lévy N vector* if it is normally distributed and if

$$\mathrm{E}W(t) = 0 \in \mathbb{R}^N, \quad \mathrm{E}\{W(s)W^T(t)\} = \int_0^s B(u)\,du \tag{2.103}$$

if $0 \leqslant s \leqslant t \leqslant T$, where

$$B(u) = Q(u)Q^T(u), \quad u \in [0, T], \tag{2.104}$$

$Q(u)$ being an $N \times N$ matrix whose elements are real-valued continuous functions of u on $[0, T]$. The matrix $B(u)$ then has the three properties mentioned above.

Unless stated otherwise, this definition for W is always implied in what follows.

The following result is a direct consequence of this discussion.

Theorem 2.34. The general Wiener-Lévy N vector $W : [0, T] \to L_2^N(\Omega)$ may be represented as in (2.94), i.e.,

$$W(t) = \int_0^t Q(u)\,dW^o(u), \quad t \in [0, T], \tag{2.105}$$

where $W^o : [0, T] \to L_2^N(\Omega)$ is the standard Wiener-Lévy N vector. □

Note that $B(u) \geqslant 0$ and not $B(u) > 0$. If $\det B(u) = 0$ at some $u \in [0, T]$, then $B(u)$ and $Q(u)$ are singular and the components of column vector "$Q(u)\,dW^o(u)$" are linearly dependent at u as functions of $\omega \in \Omega$.

We shall now derive a number of consequences of the definition of W given above without resorting to the representation given by (2.94).

a) $\mathrm{E}\{[W_m(s_2) - W_m(s_1)][W_n(t_2) - W_n(t_1)]\}$

$$= \begin{cases} 0, & \text{if } [s_1, s_2) \cap [t_1, t_2) = \emptyset \\ \int_{x_1}^{x_2} b_{mn}(u)\,du, & \text{if } [s_1, s_2) \cap [t_1, t_2) = [x_1, x_2). \end{cases} \tag{2.106}$$

This is seen, by virtue of (2.103),

$$\begin{aligned} &\mathrm{E}\{[W_m(s_2) - W_m(s_1)][W_n(t_2) - W_n(t_1)]\} \\ &\quad = \mathrm{E}\{W_m(s_2)W_n(t_2)\} - \mathrm{E}\{W_m(s_2)W_n(t_1)\} - \mathrm{E}\{W_m(s_1)W_n(t_2)\} \\ &\qquad + \mathrm{E}\{W_m(s_1)W_n(t_1)\} \\ &\quad = \left(\int_0^{s_2} - \int_0^{t_1} - \int_0^{s_1} + \int_0^{s_1}\right) b_{mn}(u)\,du = \int_{t_1}^{s_2} b_{mn}(u)\,du, \end{aligned}$$

if $s_1 \leqslant t_1 = x_1 \leqslant s_2 = x_2 \leqslant t_2$. Hence, in particular, the increments of a component process are orthogonal on disjoint intervals $[s_1, s_2)$ and $[t_1, t_2)$ or, equivalently, they are stochastically independent as the components are normally distributed and centered (Sect. 1.3.2).

However, the increments of the components are not stationary in general as

$$\mathrm{E}\{W_m(t) - W_m(s)\}^2 = \int_s^t b_{mm}(u)\, du\,, \quad \text{if} \quad s \leqslant t$$

which may not be a function of $t - s$ only.

If $B(u) = I$, the $N \times N$ identity matrix, for all $u \in [0, T]$, then $\boldsymbol{W}$ is the standard Wiener-Lévy N vector $\boldsymbol{W}^o$.

b) The components of $\boldsymbol{W}$ are of bounded variation on $[a, b] \subset [0, T]$ in the weak sense. To see this, let p be the partition of $[a, b]$ as described by (2.16). Then $p \times p$ is a partition of $[a, b]^2$ and, according to (2.106), if $C_m(x, y) = \mathrm{E}\{W_m(x)\, W_m(y)\}$,

$$\begin{aligned} V_{C_m}(p \times p) &= \sum_{i=1}^{k} \sum_{j=1}^{k} |\mathrm{E}\{[W_m(t_i) - W_m(t_{i-1})][W_m(t_j) - W_m(t_{j-1})]\}| \\ &= \sum_{i=1}^{k} \int_{t_{i-1}}^{t_i} b_{mm}(u)\, du = \int_a^b b_{mm}(u)\, du\,. \end{aligned}$$

The total variation $V_{C_m}([a, b]^2)$ is thus equal to

$$\int_a^b b_{mm}(u)\, du\,,$$

which is finite as $b_{mm}(u)$ is continuous on $[a, b]$.

c) If f and $g : [0, T] \to \mathbb{R}$ are continuous, then

$$\int_a^b f(u)\, dW_m(u) \quad \text{and} \quad \int_a^b g(v)\, dW_n(v) \tag{2.107}$$

exist and

$$\mathrm{E}\left\{\int_a^b f(u)\, dW_m(u) \int_a^b g(v)\, dW_n(v)\right\} = \int_a^b f(u)\, g(u)\, b_{mn}(u)\, du\,. \tag{2.108}$$

The existence of the integrals in (2.107) follows from (b) and from Theorem 2.30. To verify (2.108), we can compute the integrals involved, by means of Riemann-Stieltjes sums defined on the same convergent sequence $\{p_r\}_{r \in N}$ of partitions of $[a, b]$. We obtain

$$\begin{aligned} \mathrm{E}\left\{\int_a^b f(u)\, dW_m(u) \int_a^b g(v)\, dW_n(v)\right\} &= \mathrm{E}\left\{\lim_{r \to \infty} S_{f, W_m}(p_r) \lim_{q \to \infty} S_{g, W_n}(p_q)\right\} \\ &= \lim_{r \to \infty} \mathrm{E}\{S_{f, W_m}(p_r)\, S_{g, W_n}(p_r)\} \end{aligned}$$

on account of (B) in Sect. 2.1. If p_r stands for the partition described in Sect. 2.4, with the aid of (a) we get

$$\begin{aligned} &\mathrm{E}\{S_{f,W_m}(p_r)\,S_{g,W_n}(p_r)\} \\ &= \mathrm{E}\left\{\left[\sum_{i=1}^{k} f(t_i')[W_m(t_i) - W_m(t_{i-1})]\right]\left[\sum_{j=1}^{k} g(t_j')[W_n(t_j) - W_n(t_{j-1})]\right]\right\} \\ &= \sum_{i=1}^{k}\sum_{j=1}^{k} f(t_i')\,g(t_j')\,\mathrm{E}\{[W_m(t_i) - W_m(t_{i-1})][W_n(t_j) - W_n(t_{j-1})]\} \\ &= \sum_{i=1}^{k} f(t_i')\,g(t_i')\int_{t_{i-1}}^{t_i} b_{mn}(u)\,du\,. \end{aligned}$$

Since $b_{mn}(u)$ is continuous on $[t_{i-1}, t_i]$, there is a number $t_i'' \in [t_{i-1}, t_i]$ such that

$$\int_{t_{i-1}}^{t_i} b_{mn}(u)\,du = b_{mn}(t_i'')(t_i - t_{i-1})\,, \quad i = 1, \ldots, k$$

and hence we may write

$$\mathrm{E}\{S_{f,W_m}(p_r)\,S_{g,W_n}(p_r)\} = \sum_{i=1}^{k} f(t_i')\,g(t_i')\,b_{mn}(t_i'')(t_i - t_{i-1})\,.$$

As the integrals in (2.107) exist, we may choose $t_i' = t_i''$ and the right-hand side is now a Riemann sum corresponding to

$$\int_a^b f(u)\,g(u)\,b_{mn}(u)\,du\,.$$

Equation (2.108) follows as $r \to \infty$.

d) If f and $g : [0, T] \to \mathbb{R}$ are continuous and if $0 \leqslant s \leqslant t \leqslant T$, then

$$\mathrm{E}\left\{\int_0^s f(u)\,dW_m(u)\int_0^t g(v)\,dW_n(v)\right\} = \int_0^s f(u)\,g(u)\,b_{mn}(u)\,du\,. \tag{2.109}$$

This result follows from (c) using similar arguments to those used in the proof of Theorem 2.33.

We are now in a position to prove the following principal theorem.

Theorem 2.35. Let F and G be $N \times N$ matrices whose elements f_{ij} and $g_{ij} : [0, T] \to \mathbb{R}$ are continuous and let $\boldsymbol{W} : [0, T] \to L_2^N(\Omega)$ be the general Wiener-Lévy N vector. Then, with $0 \leqslant s \leqslant t \leqslant T$,

$$\mathrm{E}\left\{\int_0^s F(u)\,d\boldsymbol{W}(u)\left[\int_0^t G(v)\,d\boldsymbol{W}(v)\right]^T\right\} = \int_0^s F(u)\,B(u)\,G^T(u)\,du\,. \tag{2.110}$$

Proof. The ijth element of the left-hand side is

$$\mathrm{E}\left\{\sum_{m=1}^{N}\int_0^s f_{im}(u)\,dW_m(u)\sum_{n=1}^{N}\int_0^t g_{jn}(v)\,dW_n(v)\right\}$$

which, as shown in (d), is equal to

$$\sum_{m=1}^{N} \sum_{n=1}^{N} \int_0^s f_{im}(u)\, g_{jn}(u)\, b_{mn}(u)\, du\ .$$

It is seen as the ijth element of the right-hand side of (2.110). The proof is complete. □

In the literature one often encounters the process

$$\int_0^t G(u)\, d\boldsymbol{W}(u)\ ,$$

where $\boldsymbol{W}$ satisfies the conditions given here and $G(u)$, $u \in [0, T]$, is an $N \times N$ matrix whose elements are continuous and real functions. We observe, however, that no principal difference exists between this process and the process $\boldsymbol{W}$ itself as, according to Theorem 2.35,

$$\mathrm{E}\left\{\int_0^s G(u)\, d\boldsymbol{W}(u)\left[\int_0^t G(v)\, d\boldsymbol{W}(v)\right]^T\right\} = \int_0^s G(u)\, B(u)\, G^T(u)\, du$$

if $0 \leqslant s \leqslant t \leqslant T$, where the matrix $G(u)\, B(u)\, G^T(u)$, $u \in [0, T]$, again satisfies all the conditions in the definition of $\boldsymbol{W}$.

We also mention that within the theory of generalized functions and stochastic processes, the derivative of the general Wiener-Lévy N vector exists and is generally referred to as "white noise". It is formally

$$\dot{\boldsymbol{W}}(t) = \frac{d}{dt}\boldsymbol{W}(t)\ ; \qquad \boldsymbol{W}(t) = \int_0^t \dot{\boldsymbol{W}}(s)\, ds \quad \text{and} \tag{2.111}$$

$$\mathrm{E}\{\dot{\boldsymbol{W}}(u)\, \dot{\boldsymbol{W}}^T(v)\} = B(u)\, \delta(u-v)\ , \tag{2.112}$$

where $B(u)$ satisfies the conditions of definition for $\boldsymbol{W}$ and δ is the Dirac delta function formally defined by

$$\forall \text{ continuous } f: \mathbb{R} \to \mathbb{R},\ \int_{-\infty}^{\infty} \delta(u-v) f(u)\, du = f(v)\ .$$

2.6.2 Supplementary Exercises

Exercise 2.12. Give a formal proof of Theorem 2.35 with the aid of "white noise" as discussed above.

Exercise 2.13. Give a proof of Theorem 2.35 if $\boldsymbol{W}$ is defined as given by (2.94).

2.7 Mean-Square Calculus and Gaussian Distributions

We recall that in a closed Gaussian manifold G, linear combinations and limits in m.s. of Cauchy sequences of elements of G belong to G. Hence, difference quotients and Riemann-Stieltjes sums of elements of G belong to G, so do m.s. derivatives and integrals of processes in G.

In particular, all m.s. integrals discussed in Sect. 2.6 are Gaussian processes since they belong to the closed Gaussian manifold spanned by the (Gaussian) Wiener-Lévy process.

As seen from Sect. 1.4.3, all probability distributions associated with a Gaussian vector process may be derived from the means and the covariance function matrices.

2.8 Mean-Square Calculus and Sample Calculus

In the next chapter, it will be of interest to consider calculus of the trajectories, or sample calculus, of stochastic processes. Hence, as seen from Sect. 1.4, we shall need separable (representations of the) processes. In our discourse, all processes, starting with the Wiener-Lévy process, will be sample continuous and hence separable and will also be of second order. A natural question then arises as to whether or not a similarity exists between m.s. calculus results and those from sample calculus. The answer, as a consequence of the theorem stated below and of Theorem 2.2, is that if a second-order process has a m.s. limit as well as a limit in the sample sense, then these limits are a.s. identical. Hence, in particular, if a second-order process has a m.s. derivative or a m.s. integral as well as a derivative or an integral in the sample sense, then these derivatives or integrals are a.s. identical.

Theorem 2.36. Let $\{\Omega, \mathscr{A}, P\}$ be a probability space and let X_o and $X_n : \Omega \to \mathbb{R}$, $n \in \mathbb{N}$, be second-order random variables. If

$$X_n \xrightarrow{\text{m.s.}} X_o \tag{2.113}$$

and if

$$X_n \to X \quad \text{a.s.} \tag{2.114}$$

as $n \to \infty$, then

$$X_o = X \quad \text{a.s.} \tag{2.115}$$

Proof. The Chebyshev inequality (1.68) states that if $Y : \Omega \to \mathbb{R}$ is of second order, then, for any $\varepsilon > 0$,

$$P\{\omega : |Y| \geqslant \varepsilon\} \leqslant \frac{1}{\varepsilon^2} \mathrm{E}Y^2 . \tag{2.116}$$

Thus, (2.113) together with (2.116) lead to

$$P\{\omega : |X_o - X_n| \geqslant \varepsilon\} \leqslant \frac{1}{\varepsilon^2} \mathrm{E}(X_o - X_n)^2 \to 0 \tag{2.117}$$

as $n \to \infty$. In this case, X_n is said to converge to X_o in probability.

On the other hand, (2.114) means

$$\forall \varepsilon > 0, \ \bigcap_{m=1}^{\infty} \{\omega : |X - X_{n+m}| < \varepsilon\} \uparrow \Omega \backslash N \tag{2.118}$$

as $n \to \infty$, where N is some set with $PN = 0$. Equation (2.118) implies

$$\forall \varepsilon > 0, P\{\omega : |X - X_n| \geqslant \varepsilon\} \to 0 \tag{2.119}$$

as $n \to \infty$. Thus, X_n also converges to X in probability.

Setting

$$A = \{\omega : |X_o - X| < \varepsilon\} \quad B = \left\{\omega : |X_o - X_n| < \frac{\varepsilon}{2}\right\}$$

$$C = \left\{\omega : |X_n - X| < \frac{\varepsilon}{2}\right\}$$

then

$$B \cap C \subset A ,$$

since

$$|X_o - X| \leqslant |X_o - X_n| + |X_n - X| .$$

Hence, equivalently, $A^C \subset B^C \cup C^C$, implying $PA^C \leqslant PB^C + PC^C$, i.e.,

$$P\{\omega : |X_o - X| \geqslant \varepsilon\} \leqslant P\left\{\omega : |X_o - X_n| \geqslant \frac{\varepsilon}{2}\right\} + P\left\{\omega : |X_n - X| \geqslant \frac{\varepsilon}{2}\right\} . \tag{2.120}$$

It then follows from (2.117, 119) that

$$\forall \varepsilon > 0 ; \quad P\{\omega : |X_o - X| \geqslant \varepsilon\} = 0 .$$

Finally, setting

$$D_n = \left\{\omega : |X_o - X| \geqslant \frac{1}{n}\right\}, \qquad n \in \mathbb{N}$$

we have

$$PD_n = 0 .$$

And, as $n \to \infty$,

$$D_n \uparrow \bigcup_{i=1}^{\infty} D_i = \{\omega : |X_o - X| > 0\} ,$$

giving

$$P\{\omega : |X_o - X| > 0\} = \lim_{n\to\infty} PD_n = 0 ,$$

i.e., $X_o = X$ a.s., completing the proof. □

Example 2.5. Let $W : [0, T] \to L_2(\Omega)$ be the standard Wiener-Lévy process and let $f : [0, T] \to \mathbb{R}$ be continuously differentiable. Let us consider the m.s. as well as sample properties of the integral

$$\int_0^t f(u)\, dW(u) , \qquad t \in [0, T] . \tag{2.121}$$

According to Theorem 2.32, we immediately see that the integral in (2.121) exists in m.s. As for its sample properties, we have seen in Sect. 1.4.4 that the trajectories of W are not of bounded variation on $[0, T]$ with probability one. Hence, to determine whether the integral exists as a sample integral, we first examine

$$\int_0^t W(u)\, df(u) . \tag{2.122}$$

Being continuously differentiable, f is of bounded variation on $[0, T]$. Since the trajectories of W are continuous real functions on $[0, T]$ with probability one, (2.122) exists as a sample integral.

Finally, because

$$\int_0^t f(u)\, dW(u) = [f(u)\, W(u)]_0^t - \int_0^t W(u)\, df(u)$$

in m.s. as well as in the sample sense, it follows that the integral in (2.121) also exists in the sample sense.

We remark that although it was not used in the above, Theorem 2.23 indicates that

$$\int_0^t W(u)\,df(u) = \int_0^t W(u)f'(u)\,du$$

under these conditions.

2.8.1 Supplementary Exercise

Exercise 2.14. Consider the probability space $\{[0, 1], \mathscr{B}, \lambda\}$ (Sect 1.1) and construct on $[0, 1] \times [0, T]$ a sample differentiable process which is not m.s. differentiable at $t = 0$.

3. The Stochastic Dynamic System

In the design and analysis of a physical dynamic system, *filtering* refers to the estimation of the system state on the basis of system measurements contaminated by random noise. The Kalman-Bucy filter, being an algorithm for computing estimates of the state vector, deals with a *stochastic* dynamic system driven by forces whose random components are modeled by Brownian motion. In this chapter we are concerned with this system. Since only a sample of the stochastic processes is realized at the end of each physical experiment modeled by the dynamic system, the use of sample calculus is appropriate in the mathematical model.

As we shall see, however, the mathematical treatment can also be carried out by means of m.s. calculus. Theorem 2.36 (Sect. 2.8) is applicable and hence the sample derivatives and integrals we shall meet also exist in the m.s. sense, and are a.s. identical to the corresponding sample results. This enables us to calculate means, variances and correlations. Moreover, all stochastic processes encountered in the development turn out to belong to the same Gaussian manifold so that all distribution functions and relevant statistics may be computed (Sect. 2.7).

3.1 System Description

In basic Kalman-Bucy filtering, the dynamic system is assumed to be a linear stochastic system whose state N vector $\boldsymbol{X}$ satisfies a system of linear stochastic differential equations of the type

$$\frac{d}{dt}\boldsymbol{X}(t) = A(t)\boldsymbol{X}(t) + \boldsymbol{g}(t) + \dot{\boldsymbol{W}}(t) , \quad t \in [0, T] \tag{3.1}$$

with random initial condition

$$\boldsymbol{X}(0) = \boldsymbol{C} . \tag{3.2}$$

In (3.1), the stochastic input function or "noise" $\dot{\boldsymbol{W}}(t)$ is the generalized derivative of the Wiener-Lévy N vector (Sect. 2.6). However, since this derivative does not exist in the usual sense, and to avoid the use of generalized stochastic calculus, we consider instead the system described below:

$$X(t) = C + \int_0^t A(s)X(s)\,ds + f(t) + W(t)\,, \quad t \in [0, T]\,, \tag{3.3}$$

where $W: [0, T] \rightarrow L_2^N(\Omega)$ is the Wiener-Lévy N vector, $C \in L_2^N(\Omega)$ is a prescribed, normally distributed N vector, stochastically independent of W, on $[0, T]$, and A and f are, respectively, an $N \times N$ matrix and an N vector whose elements and components are ordinary, continuous real functions defined on $[0, T]$.

As indicated earlier, we are interested in the sample solutions, as well as solutions in the m.s. sense, to the system described by (3.3). Since the trajectories of the components of W and the elements and components of A and f are continuous on $[0, T]$, well-known existence and uniqueness theorems for solutions of ordinary (linear) differential equations immediately imply that (3.3) has a unique and continuous sample solution. In the next section, we show that (3.3) also has a unique and continuous solution in the m.s. sense (see, e.g., [3.1] or the first sections of [3.2]).

We remark that many results derived in $L_2^N(\Omega)$ are also valid and meaningful in $\mathbb{R}^N$, the Cartesian product of degenerated spaces $L_2(\Omega)$ where Ω consists of one point only. The theorem of existence and uniqueness of the m.s. solution to (3.3), for example, is one of them, which implicitly also gives the uniqueness and existence proof of continuous solutions to systems of ordinary linear differential equations.

Unless stated otherwise, all calculus used in what follows means m.s. calculus.

3.2 Uniqueness and Existence of m.s. Solution to (3.3)

3.2.1 The Banach Space $L_2^N(\Omega)$

We have thus far used $L_2^N(\Omega)$ simply as a Cartesian product of N identical spaces $L_2(\Omega)$; for the following development, a topology in $L_2^N(\Omega)$ is also needed. We introduce it by means of a norm constructed as follows. For all X in $L_2^N(\Omega)$, we define

$$\|X\|_N = \|X_1\| + \ldots + \|X_N\|\,, \tag{3.4}$$

where $\|X_i\|$ is the norm of X_i as defined in (1.59), i.e.,

$$\|X_i\| = (\mathrm{E}X_i^2)^{1/2}\,. \tag{3.5}$$

We see that $\|X\|_N$ is indeed a norm since

a) $\|cX\|_N = |c| \, \|X\|_N$ for all $c \in \mathbb{R}$ and X in $L_2^N(\Omega)$

b) $\|X + Y\|_N \leqslant \|X\|_N + \|Y\|_N$ for all X and Y in $L_2^N(\Omega)$

c) $\|X\|_N \geqslant 0$; $\|X\|_N = 0$ only if $X_i = 0$ a.s. for all i.

In fact, as seen from Sect. 1.4.1, $\|X\|_N$ is a seminorm but leads to a norm on the linear space of equivalence classes of a.s. identical random N vectors. However, for convenience, no distinction will be made between random N vectors and their classes and no special notations used for equivalence classes and operations performed on them.

Associated with the above norm, $L_2^N(\Omega)$ is a complete space and hence a Banach space, as will be shown below. Let $\{X^{(n)}\}_{n \in N}$, $X^{(n)} \in L_2^N(\Omega)$, be a Cauchy sequence, i.e., let

$$\|X^{(m)} - X^{(n)}\|_N \to 0 \quad \text{as} \quad m, n \to \infty$$

Then,

$$\|X_i^{(m)} - X_i^{(n)}\| \to 0$$

for all i. Hence, for each index i, the sequence $\{X_i^{(n)}\}_{n \in N}$ is a Cauchy sequence in the complete space $L_2(\Omega)$ with a unique m.s. limit, say, X_i. It then follows that

$$X^{(n)} = \begin{bmatrix} X_1^{(n)} \\ \vdots \\ X_N^{(n)} \end{bmatrix} \to \begin{bmatrix} X_1 \\ \vdots \\ X_N \end{bmatrix}$$

in the sense that

$$\sum_{i=1}^{N} \|X_i^{(n)} - X_i\| \to 0$$

showing the completeness of $L_2^N(\Omega)$ under the norm defined above. Unless stated otherwise, $L_2^N(\Omega)$ shall mean the Banach space defined above.

Some useful relationships associated with the norm defined by (3.4) are noted below.

Consider a real-valued $N \times N$ matrix $A = [a_{ij}]$, $i, j = 1, 2, \ldots, N$. Defining

$$|A| = \sum_{i,j=1}^{N} |a_{ij}| \,, \quad \text{then} \tag{3.6}$$

$$\|AX\|_N = \left\| \begin{bmatrix} a_{11}X_1 + \ldots + a_{1N}X_N \\ \vdots \\ a_{N1}X_1 + \ldots + a_{NN}X_N \end{bmatrix} \right\|_N$$

$$= \|a_{11}X_1 + \ldots + a_{1N}X_N\| + \ldots + \|a_{N1}X_1 + \ldots + a_{NN}X_N\|$$

$$\leqslant \left\{ \sum_{i,j=1}^{N} |a_{ij}| \right\} \{\|X_1\| + \ldots + \|X_N\|\} = |A| \, \|X\|_N \, . \tag{3.7}$$

If the components X_i of X are m.s. continuous mappings of $[0, T]$ into $L_2(\Omega)$, it is then seen from Sect. 2.2 that $\|X\|_N$ is a continuous function of t on $[0, T]$. Thus, for each $t \in [0, T]$, Theorem 2.22 shows that

$$\left\| \int_0^t X(s)\,ds \right\|_N = \left\| \int_0^t X_1(s)\,ds \right\| + \ldots + \left\| \int_0^t X_N(s)\,ds \right\|$$

$$\leqslant \int_0^t \|X_1(s)\|\,ds + \ldots + \int_0^t \|X_N(s)\|\,ds$$

$$= \int_0^t \|X(s)\|_N\,ds \, . \tag{3.8}$$

Moreover, if each element of A is a continuous and real function of s on $[0, T]$, then so is $|A(s)|$ and, for each $t \in [0, T]$, we have

$$\left\| \int_0^t A(s)X(s)\,ds \right\|_N \leqslant \int_0^t \|A(s)X(s)\|_N\,ds$$

$$\leqslant \int_0^t |A(s)| \, \|X(s)\|_N\,ds \, . \tag{3.9}$$

3.2.2 Uniqueness

The theorem given below shows that if a m.s. continuous solution to (3.3) exists, it is unique.

Theorem 3.1. Suppose that both X and $Y : [0, T] \to L_2^N(\Omega)$ are m.s. continuous on $[0, T]$ and satisfy (3.3). Then

$$X(t) = Y(t) \text{ a.s.} \tag{3.10}$$

for all $t \in [0, T]$.

Proof. Let $X(t)$ and $Y(t)$ be two N vector processes satisfying (3.3). Then

$$X(t) - Y(t) = \int_0^t A(s)[X(s) - Y(s)]\,ds \, , \quad t \in [0, T]$$

or, putting $\mathbf{Z}(t) = \mathbf{X}(t) - \mathbf{Y}(t)$,

$$\mathbf{Z}(t) = \int_0^t A(s)\mathbf{Z}(s)\,ds\,. \tag{3.11}$$

It then follows from (3.9) that

$$\|\mathbf{Z}(t)\|_N \leqslant \int_0^t |A(s)|\,\|\mathbf{Z}(s)\|_N\,ds\,. \tag{3.12}$$

We recall that $|A|$ and $\|\mathbf{Z}\|_N$ are continuous real functions on $[0, T]$. Let us define

$$M = \max_{t\in[0,T]} |A(t)|\,, \qquad m = \max_{t\in[0,T]} \|\mathbf{Z}(t)\|_N\,.$$

Equation (3.12) gives

$$\|\mathbf{Z}(t)\|_N \leqslant \int_0^t Mm\,ds = Mmt$$

and hence, again according to (3.12),

$$\|\mathbf{Z}(t)\|_N \leqslant \int_0^t M(Mms)\,ds = m\,\frac{(Mt)^2}{2}$$

$$\|\mathbf{Z}(t)\|_N \leqslant \int_0^t M\left(m\,\frac{(Ms)^2}{2}\right)ds = m\,\frac{(Mt)^3}{3!}\,.$$

By induction, we have

$$\|\mathbf{Z}(t)\|_N \leqslant m\,\frac{(Mt)^n}{n!} \tag{3.13}$$

for each natural value of n. Hence, since $(Mt)^n/n! \to 0$ as $n \to \infty$,

$$\|\mathbf{Z}(t)\|_N = 0$$

and $\mathbf{Z}(t) = \mathbf{0}$ a.s. for all $t \in [0, T]$. □

3.2.3 The Homogeneous System

Concerning the existence of the m.s. solution to (3.3) and its construction, we consider first the homogeneous system. We shall need the following theorem.

Theorem 3.2. Suppose that the processes $X^{(n)} : [0, T] \to L_2(\Omega)$ are m.s. continuous for all $n \in \mathbb{N}$ and that $X^{(n)}(t) \to X(t)$ in m.s. as $n \to \infty$, uniformly in $t \in [0, T]$. Then

a) X also maps $[0, T]$ into $L_2(\Omega)$

b) X is m.s. continuous on $[0, T]$

c) $\lim_{n\to\infty} \int_0^T X^{(n)}(t)\,dt = \int_0^T X(t)\,dt$.

Proof. Assertion (a) is implied by the completeness of $L_2(\Omega)$. As for (b), for any t and $t + h$ in $[0, T]$,

$$\|X(t+h) - X(t)\| \leqslant \|X(t+h) - X^{(n)}(t+h)\| + \|X^{(n)}(t+h) - X^{(n)}(t)\| + \|X^{(n)}(t) - X(t)\| . \tag{3.14}$$

Given $\varepsilon > 0$ and using uniform convergence, we first find an N such that

$$\|X(s) - X^{(n)}(s)\| < \frac{\varepsilon}{3}$$

for all $n > N$ and all s in $[0, T]$. Now, fix $n \in N$ and find a $\delta > 0$ such that

$$\|X^{(n)}(t+h) - X^{(n)}(t)\| < \frac{\varepsilon}{3}$$

for all h with $|h| < \delta$. Equation (3.13) thus gives

$$\|X(t+h) - X(t)\| < \varepsilon$$

if $|h| < \delta$, showing m.s. continuity of X on $[0, T]$. Assertion (c) follows because, given $\varepsilon > 0$ and again using uniform convergence, there is an N' such that for all $n > N'$

$$\|X(t) - X^{(n)}(t)\| < \varepsilon \tag{3.15}$$

for all $t \in [0, T]$. Now, the integrals in assertion (c) exist because of (b) and, in view of (3.15),

$$\left\| \int_0^T X(t)\,dt - \int_0^T X^{(n)}(t)\,dt \right\| \leqslant \int_0^T \|X(t) - X^{(n)}(t)\|\,dt \leqslant \varepsilon T$$

if $n > N'$. □

Consider now the homogeneous part of the dynamic system defined by (3.3), i.e.,

$$\boldsymbol{X}(t) = \boldsymbol{C} + \int_0^t A(s)\boldsymbol{X}(s)\,ds\,, \quad t \in [0, T] \tag{3.16}$$

and construct a continuous solution to it in the m.s. sense. Let $\boldsymbol{X}^{(o)} : [0, T] \to L_2^N(\Omega)$ be an arbitrary N vector with m.s. continuous components and define $\mathscr{A}\boldsymbol{X}^{(o)}$ as

$$(\mathscr{A}\boldsymbol{X}^{(o)})(t) = \int_0^t A(s)\boldsymbol{X}^{(o)}(s)\,ds\,, \quad t \in [0, T] . \tag{3.17}$$

Then

$$\mathscr{A}X^{(o)} : [0, T] \to L_2^N(\Omega) \tag{3.18}$$

and it has m.s. continuous components. From (3.12), we have

$$\|(\mathscr{A}X^{(o)})(t)\|_N \leqslant \int_0^T |A(s)| \, \|X^{(o)}(s)\|_N ds \leqslant \int_0^t Mm \, ds = Mmt ,$$

where

$$M = \max_{t \in [0,T]} |A(t)| , \qquad m = \max_{t \in [0,T]} \|X^{(o)}(t)\|_N .$$

Similarly, $\mathscr{A}^2 X^{(o)}$ is defined by

$$(\mathscr{A}^2 X^{(o)})(t) = \int_0^t A(s)[(\mathscr{A}X^{(o)})(s)] \, ds .$$

Hence

$$\mathscr{A}^2 X^{(o)} : [0, T] \to L_2^N(\Omega)$$

and it has m.s. continuous components. Furthermore,

$$\|\mathscr{A}^2 X^{(o)}(t)\|_N \leqslant \int_0^t |A(s)| \, \|(\mathscr{A}X^{(o)})(s)\|_N ds$$

$$\leqslant \int_0^t M(Mms) \, ds = m \frac{(Mt)^2}{2}$$

and, by induction, $\mathscr{A}^n X^{(o)} : [0, T] \to L_2^N(\Omega)$ has m.s. continuous components and (Sect. 3.2.2)

$$\|(\mathscr{A}^n X^{(o)})(t)\|_N \leqslant m \frac{(Mt)^n}{n!} , \qquad n \in \mathbb{N} . \tag{3.19}$$

We remark that an analogous result is obtained if $X^{(o)}$ is replaced by the time-independent N vector C.

If both Y and $Z : [0, T] \to L_2^N(\Omega)$ have m.s. continuous components, $\mathscr{A}Y$, $\mathscr{A}Z$ and $\mathscr{A}(Y + Z)$ are defined; they map $[0, T]$ into $L_2^N(\Omega)$ and also have m.s. continuous components. It is easily seen that

$$\mathscr{A}(Y + Z) = \mathscr{A}Y + \mathscr{A}Z . \tag{3.20}$$

Now, we are in a position to define recursively

$$X^{(n)} = C + \mathscr{A}X^{(n-1)} , \qquad n \in \mathbb{N} , \tag{3.21}$$

giving

$$
\begin{aligned}
\boldsymbol{X}^{(1)} &= \boldsymbol{C} + \mathscr{A}\boldsymbol{X}^{(o)} \\
\boldsymbol{X}^{(2)} &= \boldsymbol{C} + \mathscr{A}\boldsymbol{X}^{(1)} = \boldsymbol{C} + \mathscr{A}\boldsymbol{C} + \mathscr{A}^2\boldsymbol{X}^{(o)} \\
&\vdots \\
\boldsymbol{X}^{(n)} &= \boldsymbol{C} + \mathscr{A}\boldsymbol{X}^{(n-1)} = \boldsymbol{C} + \ldots + \mathscr{A}^{n-1}\boldsymbol{C} + \mathscr{A}^n\boldsymbol{X}^{(o)} .
\end{aligned}
$$

With the results presented above, we have

$$\boldsymbol{X}^{(n)} : [0, T] \to L_2^N(\Omega)$$

and it has m.s. continuous components. Also, for any $k \in \mathbb{N}$,

$$\boldsymbol{X}^{(n+k)} - \boldsymbol{X}^{(n)} = \mathscr{A}^n\boldsymbol{C} + \ldots + \mathscr{A}^{n+k-1}\boldsymbol{C} + \mathscr{A}^{n+k}\boldsymbol{X}^{(o)} - \mathscr{A}^n\boldsymbol{X}^{(o)} .$$

Hence, if $\|\boldsymbol{C}\|_n = c$, (3.19) leads to

$$
\begin{aligned}
&\|\boldsymbol{X}^{(n+k)}(t) - \boldsymbol{X}^{(n)}(t)\|_N \\
&\quad \leqslant \|(\mathscr{A}^n\boldsymbol{C})(t)\|_n + \ldots + \|(\mathscr{A}^{n+k-1}\boldsymbol{C})(t)\|_N + \|(\mathscr{A}^{n+k}\boldsymbol{X}^{(o)})(t)\|_N \\
&\qquad + \|(\mathscr{A}^n\boldsymbol{X}^{(o)})(t)\|_N \\
&\quad \leqslant c\frac{(Mt)^n}{n!} + \ldots + c\frac{(Mt)^{n+k-1}}{(n+k-1)!} + m\frac{(Mt)^{n+k}}{(n+k)!} + m\frac{(Mt)^n}{n!} \to 0
\end{aligned}
$$

as $n \to \infty$, uniformly in $t \in [0, T]$. Thus, $\{\boldsymbol{X}^{(n)}(t)\}_{n \in \mathbb{N}}$ is a Cauchy sequence in $L_2^N(\Omega)$ for all $t \in [0, T]$ and a limit $\boldsymbol{X}(t)$ exists such that

$$\boldsymbol{X} : [0, T] \to L_2^N(\Omega) \quad \text{and}$$

$$\|\boldsymbol{X}^{(n)}(t) - \boldsymbol{X}(t)\|_N \to 0 \tag{3.22}$$

as $n \to \infty$. As a consequence of these results, this convergence is also uniform in $t \in [0, T]$.

Concerning the components, for each $i = 1, \ldots, N$, we have

$$X_i^{(n)}(t) \to X_i(t) \quad \text{in} \quad \text{m.s.}$$

as $n \to \infty$, uniformly in $t \in [0, T]$. Since the elements $X_i^{(n)}(t)$ are m.s. continuous on $[0, T]$, Theorem 3.2 shows that the limit $X_i(t)$ is also m.s. continuous on $[0, T]$.

Relation (3.21) may be written in component form as

$$X_i^{(n)}(t) = C_i + \left[\int_0^t a_{i1}(s)X_1^{(n-1)}(s)\,ds + \ldots + \int_0^t a_{iN}(s)X_N^{(n-1)}(s)\,ds\right] . \tag{3.23}$$

Again following Theorem 3.2, as $n \to \infty$ (3.23) gives

$$X_i(t) = C_i + \int_0^t a_{i1}(s) X_1(s)\, ds + \ldots + \int_0^t a_{iN}(s) X_N(s)\, ds\,, \quad t \in [0, T] \tag{3.24}$$

for all $i = 1, \ldots, N$. Hence

$$\boldsymbol{X}(t) = \boldsymbol{C} + \int_0^t A(s) \boldsymbol{X}(s)\, ds \tag{3.25}$$

and, as constructed above, $\boldsymbol{X} : [0, T] \to L_2^N(\Omega)$ is a solution to (3.16) in the m.s. sense, of which the components are m.s. continuous.

Combining this result with Theorem 3.1 we have the following result.

Theorem 3.3. The homogeneous system of (3.3) has a unique and continuous solution in the m.s. sense.

Theorem 3.4. The unique m.s. continuous solution to the homogeneous system of (3.3) may be represented by

$$\boldsymbol{X}(t) = \Phi(t) \boldsymbol{C}, \quad t \in [0, T]\,, \quad \text{where} \tag{3.26}$$

$$\frac{d}{dt} \Phi(t) = A(t) \Phi(t) \quad \text{and} \tag{3.27}$$

$$\Phi(0) = I, \text{ the } N \times N \text{ identity matrix.} \tag{3.28}$$

Proof. Consider first the degenerate case in which $\boldsymbol{C} = \boldsymbol{c}$, a deterministic real constant N vector. The homogeneous system (3.16) becomes

$$\frac{d}{dt} \boldsymbol{X}(t) = A(t) \boldsymbol{X}(t)\,, \quad t \in [0, T]$$

$$\boldsymbol{X}(0) = \boldsymbol{c} \tag{3.29}$$

or, equivalently,

$$\boldsymbol{X}(t) = \boldsymbol{c} + \int_0^t A(s) \boldsymbol{X}(s)\, ds\,, \quad t \in [0, T]\,. \tag{3.30}$$

Letting

$$\boldsymbol{c} = \begin{bmatrix} 1 \\ 0 \\ \vdots \\ \vdots \\ 0 \end{bmatrix}, \begin{bmatrix} 0 \\ 1 \\ 0 \\ \vdots \\ 0 \end{bmatrix}, \ldots, \begin{bmatrix} 0 \\ \vdots \\ \vdots \\ 0 \\ 1 \end{bmatrix}$$

successively, we obtain N unique solutions to (3.29) since Theorem 3.3 applies in ordinary real analysis as a special case as well. Combining these results, the $N \times N$ matrix solution $\Phi(t)$ of (3.29) with the $N \times N$ identity matrix as initial condition satisfies

$$\frac{d}{dt}\Phi(t) = A(t)\Phi(t), \quad t \in [0, T]$$

$$\Phi(0) = I \tag{3.31}$$

or, equivalently,

$$\Phi(t) = I + \int_0^t A(s)\Phi(s)\,ds. \tag{3.32}$$

This is a well-known result in ordinary real analysis. The matrix Φ is called the *fundamental solution* to (3.29 or 30) and it is invertible at any $t \in [0, T]$.

Let us now return to (3.16) and rewrite it as

$$\boldsymbol{X}(t) - \int_0^t A(s)\boldsymbol{X}(s)\,ds = \boldsymbol{C}, \quad t \in [0, T]. \tag{3.33}$$

Now,

$$\boldsymbol{X}(t) = \Phi(t)\boldsymbol{C}, \quad t \in [0, T] \tag{3.34}$$

represents a solution to (3.33) since, on account of (3.31), the substitution of (3.34) into the left-hand side of (3.33) gives

$$\begin{aligned}\Phi(t)\boldsymbol{C} - \int_0^t A(s)\Phi(s)\boldsymbol{C}ds &= \Phi(t)\boldsymbol{C} - \int_0^t \frac{d}{ds}\Phi(s)\boldsymbol{C}\,ds \\ &= \Phi(t)\boldsymbol{C} - \Phi(t)\boldsymbol{C} + \Phi(0)\boldsymbol{C} \\ &= I\boldsymbol{C} = \boldsymbol{C}.\end{aligned}$$

This completes the proof. □

3.2.4 The Inhomogeneous System

Consider now the system represented by (3.3) and rewrite it in the form

$$\boldsymbol{X}(t) - \int_0^t A(s)\boldsymbol{X}(s)\,ds = \boldsymbol{C} + \boldsymbol{f}(t) + \boldsymbol{W}(t). \tag{3.35}$$

In the preceding section, we have seen that the substitution of

$$\boldsymbol{X}(t) = \Phi(t)\boldsymbol{C}$$

into the left-hand side of (3.33) yields $\boldsymbol{C}$, where $\boldsymbol{C}$ is a time-independent stochastic N vector. Now, we shall assume

$$\boldsymbol{C} : [0, T] \to L_2^N(\Omega)$$

with m.s. continuous components and, applying the "method of variation of a constant", find a construction of $\boldsymbol{C}$ such that

$$\boldsymbol{X}(t) = \Phi(t)\boldsymbol{C}(t) \tag{3.36}$$

is a solution to (3.35).

Using (3.31), substituting (3.36) into the left-hand side of (3.35) gives

$$\begin{aligned}
&\Phi(t)\boldsymbol{C}(t) - \int_0^t A(s)\Phi(s)\boldsymbol{C}(s)\,ds \\
&\quad = \Phi(t)\boldsymbol{C}(t) - \int_0^t \left[\frac{d}{ds}\Phi(s)\right]\boldsymbol{C}(s)\,ds \\
&\quad = \Phi(t)\boldsymbol{C}(t) - [\Phi(s)\boldsymbol{C}(s)]_0^t + \int_0^t \Phi(s)\,d\boldsymbol{C}(s) \\
&\quad = \boldsymbol{C}(0) + \int_0^t \Phi(s)\,d\boldsymbol{C}(s)\,.
\end{aligned}$$

For this result to be identical to the right-hand side of (3.35), we must have

$$\boldsymbol{C}(0) + \int_0^t \Phi(s)\,d\boldsymbol{C}(s) = \boldsymbol{C} + \boldsymbol{f}(t) + \boldsymbol{W}(t)\,, \qquad t \in [0, T]\,, \tag{3.37}$$

leading to

$$\boldsymbol{C}(s) = \boldsymbol{C} + \boldsymbol{f}(0) + \int_0^s \Phi^{-1}(r)\,d\boldsymbol{f}(r) + \int_0^s \Phi^{-1}(r)\,d\boldsymbol{W}(r)\,. \tag{3.38}$$

This can be verified by noting that, from (3.37),

$$\boldsymbol{C}(0) = \boldsymbol{C} + \boldsymbol{f}(0)$$

and, following Theorem 2.31,

$$\begin{aligned}
\int_0^t \Phi(s)\,d\boldsymbol{C}(s) &= \boldsymbol{0} + \int_0^t \Phi(s)\,d\left[\int_0^s \Phi^{-1}(r)\,d\boldsymbol{f}(r)\right] \\
&\quad + \int_0^t \Phi(s)\,d\left[\int_0^s \Phi^{-1}(r)\,d\boldsymbol{W}(r)\right] \\
&= \int_0^t d\boldsymbol{f}(s) + \int_0^t d\boldsymbol{W}(s) = \boldsymbol{f}(t) - \boldsymbol{f}(0) + \boldsymbol{W}(t)\,.
\end{aligned} \tag{3.39}$$

We have thus shown that the (unique) m.s. solution to (3.35) may be represented by

$$X(t) = \Phi(t)\left[C + f(0) + \int_0^t \Phi^{-1}(s)\,df(s) + \int_0^t \Phi^{-1}(s)\,dW(s)\right],$$
$$t \in [0, T]\,. \tag{3.40}$$

It is remarked that this representation is also meaningful in the sample sense, i.e., it represents the a.s. unique sample solution to (3.35) following from the application of the arguments in Example 2.5 to the integrals in (3.38–40).

A number of important properties associated with (3.40) are noted below.

a) Following Sect. 2.6,

$$\Phi(t)\int_0^t \Phi^{-1}(s)\,dW(s)$$

is normally distributed. Since C is also Gaussian and C and $W(t)$, $t \in [0, T]$, are stochastically independent, $X(t)$, $t \in [0, T]$, is also normally distributed.

b) Applying Theorem 2.20 to the dynamic system defined by (3.3) and to solution (3.40), since $\mathrm{E}W(t) = 0$, the expectation $\mathrm{E}X(t)$ satisfies

$$\mathrm{E}X(t) = \mathrm{E}C + \int_0^t A(s)\,\mathrm{E}X(s)\,ds + f(t)\,, \quad t \in [0, T] \tag{3.41}$$

with the solution

$$\mathrm{E}X(t) = \Phi(t)\left[\mathrm{E}C + f(0) + \int_0^t \Phi^{-1}(s)\,df(s)\right]. \tag{3.42}$$

c) Letting

$$X_o(t) = X(t) - \mathrm{E}X(t)\,, \quad t \in [0, T] \tag{3.43}$$

$$C_o = C - \mathrm{E}C \tag{3.44}$$

the system (3.3) and its solution (3.40) may be decomposed into a deterministic part and a stochastic part, the deterministic part being (3.41, 42). The stochastic part consists of

$$X_o(t) = C_o + \int_0^t A(s)X_o(s)\,ds + W(t)\,, \quad t \in [0, T]\,, \tag{3.45}$$

with the solution

$$X_o(t) = \Phi(t)\left[C_o + \int_0^t \Phi^{-1}(s)\,dW(s)\right] \tag{3.46}$$

as is seen by substracting (3.41) from (3.3), and (3.42) from (3.40).

d) According to Theorem 2.35, we have

$$\mathrm{E}X_o(0) = 0 \tag{3.47}$$

$$\mathrm{E}\{X_o(s)X_o^T(t)\} = \Phi(s)\left[\mathrm{E}\{C_oC_o^T\} + \int_0^s \Phi^{-1}(u)B(u)[\Phi^{-1}(u)]^T du\right]\Phi^T(t)\,, \tag{3.48}$$

if $0 \leqslant s \leqslant t \leqslant T$, where $B(u)$ is defined in (2.98). Let

$$V(t) = \mathrm{E}\{X_o(t)X_o^T(t)\}\,. \tag{3.49}$$

Then, because of (3.31), it satisfies

$$\begin{aligned}\frac{d}{dt}V(t) &= A(t)V(t) + \Phi(t)\{\Phi^{-1}(t)B(t)[\Phi^{-1}(t)]^T\}\Phi^T(t) + V(t)A^T(t)\\ &= A(t)V(t) + B(t) + V(t)A^T(t) \quad \text{and}\\ V(0) &= \mathrm{E}\{C_oC_o^T\}\,.\end{aligned} \tag{3.50}$$

The results given above are formalized in a theorem stated below.

Theorem 3.5. Let the dynamic system be represented by

$$X(t) = C + \int_0^t A(s)X(s)\,ds + W(t)\,, \quad t \in [0, T]\,, \tag{3.51}$$

where $A(s)$, $s \in [0, T]$, is a deterministic $N \times N$ matrix with real-valued continuous components; C is a normally distributed N vector with $\mathrm{E}C = 0$; $W(t)$, $t \in [0, T]$, is the Wiener-Lévy N vector with $\mathrm{E}W(t) = 0$ and $\mathrm{E}\{W(s)W^T(t)\} = \int_0^s B(u)\,du$ if $0 \leqslant s \leqslant t \leqslant T$, where $B(u)$, $u \in [0, T]$, is symmetric and semidefinite positive at each u and whose elements are continuous real-valued functions and so are the samples of $W(t)$; and C and $W(t)$, $t \in [0, T]$, are stochastically independent.

Then, there is a (a.s.) unique solution, both in the m.s. and in the sample sense to (3.51). In both senses, it may be represented by

$$X(t) = \Phi(t)\left[C + \int_0^t \Phi^{-1}(s)\,dW(s)\right], \quad \text{where} \tag{3.52}$$

$$\left.\begin{aligned}&X(t) \text{ is normally distributed;}\\ &\mathrm{E}X(t) = 0;\\ &\mathrm{E}\{X(s)X^T(t)\} = \Phi(s)\Big[\mathrm{E}\{CC^T\}\\ &\qquad + \int_0^s \Phi^{-1}(u)B(u)[\Phi^{-1}(u)]^T du\Big]\Phi^T(t)\end{aligned}\right\} \tag{3.53}$$

if $0 \leqslant s \leqslant t \leqslant T$. And $V(t) = \mathrm{E}\{X(t)X^T(t)\}$ satisfies

$$\left.\begin{aligned}&\frac{d}{dt}V(t) = A(t)V(t) + B(t) + V(t)A^T(t)\,, \quad t \in [0, T]\\ &V(0) = \mathrm{E}\{CC^T\}\,.\end{aligned}\right\} \tag{3.54}$$

Without proof, we mention that $\boldsymbol{X}(t)$ is a Markov vector process as well as a Martingale, and is sometimes called a Gaussian-Markov vector process. However, we shall not use these properties explicitly since the statistical properties of $\boldsymbol{X}(t)$ are completely determined by (3.53).

Finally, we remark that a commonly used notation for system (3.51) is

$$\left.\begin{aligned} d\boldsymbol{X}(t) &= A(t)\boldsymbol{X}(t) + d\boldsymbol{W}(t)\,, \quad t \in [0, T] \\ \boldsymbol{X}(0) &= \boldsymbol{C}\,. \end{aligned}\right\} \tag{3.55}$$

3.2.5 Supplementary Exercises

Exercise 3.1. There are many ways of defining a norm in $L_2^N(\Omega)$. It can be done, for example, by means of an inner product. Show that if $\boldsymbol{X}$ and $\boldsymbol{Y}$ are elements of $L_2^N(\Omega)$ and if M is a fixed symmetric positive definite $N \times N$ matrix with real-valued elements, then

$$(\boldsymbol{X}, \boldsymbol{Y})_N = \mathrm{E}\{\boldsymbol{X}^T M \boldsymbol{Y}\}$$

is an inner product in $L_2^N(\Omega)$.

Exercise 3.2. Consider the system described by

$$\left.\begin{aligned} \frac{d}{dt}\boldsymbol{X}(t) &= A(t)\boldsymbol{X}(t) + \boldsymbol{G}(t)\,, \quad t \in [0, T] \\ \boldsymbol{X}(0) &= \boldsymbol{C}\,, \end{aligned}\right\} \tag{3.56}$$

where $A(t)$ is a deterministic $N \times N$ matrix with real-valued continuous components, $\boldsymbol{C} \in L_2^N(\Omega)$, and $\boldsymbol{G} : [0, T] \to L_2^N(\Omega)$ with m.s. continuous real-valued components. Show that there is a unique m.s. solution to (3.56) and that it may be represented by

$$\boldsymbol{X}(t) = \boldsymbol{\Phi}(t)\left[\boldsymbol{C} + \int_0^t \boldsymbol{\Phi}^{-1}(s)\boldsymbol{G}(s)\,ds\right], \tag{3.57}$$

where $\boldsymbol{\Phi}$ is the fundamental solution to the homogeneous part of (3.56).

Exercise 3.3. Consider (3.56) again with $\boldsymbol{G}(t)$ replaced by the white noise $\dot{\boldsymbol{W}}(t)$, the generalized derivative of the Wiener-Lévy N vector $\boldsymbol{W}(t)$. Using the result of Exercise 3.2, show formally that

$$\boldsymbol{X}(t) = \boldsymbol{\Phi}(t)\left[\boldsymbol{C} + \int_0^t \boldsymbol{\Phi}^{-1}(s)\,d\boldsymbol{W}(t)\right] \tag{3.58}$$

is a solution representation of the system.

3.3 A Discussion of System Representation

In the context of Kalman-Bucy filtering, it is instructive to consider the scope and limitations of using (3.3) as the mathematical model for a physical experiment. The use of the Wiener-Lévy N vector as a forcing term is our primary concern since it possesses some undesirable properties, such as the nondifferentiability of the samples of its components.

In what follows, to justify the use of $\boldsymbol{W}(t)$ in the mathematical model, we shall show that there are stochastic processes with smooth and hence realistic samples which approximate arbitrarily closely the sample functions of the components of $\boldsymbol{W}(t)$, with responses approximating the solution of (3.3). It is of interest to note that there are situations where such a justification fails [3.3]. Because of its exceptional computational properties (e.g., Theorem 2.35), the Wiener-Lévy N vector is retained in the model.

There are a number of ways of approximating $\boldsymbol{W}(t)$ on $[0, T]$ in the sense stated above [3.4]. We choose here to work with convolutions with testing functions as used in the theory of generalized functions [3.5].

Let $p : \mathbb{R} \to \mathbb{R}$ have the following properties:

a) p is nonnegative and symmetric with compact support (i.e., $p(t) = 0$ outside a closed and bounded subset of $\mathbb{R}$);

b) p is smooth (infinitely often differentiable) on $\mathbb{R}$;

c) $\int_{-\infty}^{\infty} p(t)\, dt = 1$.

Given p, we define $p_n : \mathbb{R} \to \mathbb{R}$, $n \in \mathbb{N}$, as

$$p_n(t) = np(nt) . \tag{3.59}$$

It is seen that p_n also enjoys properties (a–c) given above. As an example, $p_n(t)$ can take the form (Fig. 3.1)

$$\begin{aligned} p_n(t) &= n\mathrm{e}^{1/(n^2t^2-1)} \Big/ \int_{-1}^{1} \mathrm{e}^{1/(s^2-1)}\, ds, && |t| < \frac{1}{n} \\ &= 0 , && |t| \geqslant \frac{1}{n} . \end{aligned}$$

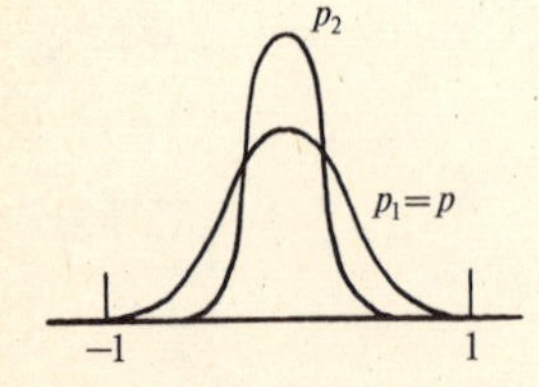

Fig. 3.1. Graphs of $p_1(t)$ and $p_2(t)$

For each p_n let us define $q_n : \mathbb{R} \to \mathbb{R}$ as

$$q_n(t) = \int_{-\infty}^{t} p_n(s)\, ds \,. \tag{3.60}$$

It is seen that q_n converges pointwise as $n \to \infty$ to the unit-step Heaviside function h, i.e.,

$$\lim_{n\to\infty} q_n(t) = h(t) \,, \quad t \in [0, T] \,, \quad \text{where} \tag{3.61}$$

$$\begin{aligned} h(t) &= 0 \,, \quad t < 0 \\ &= \tfrac{1}{2} \,, \quad t = 0 \\ &= 1 \,, \quad t > 0 \,. \end{aligned} \tag{3.62}$$

Let $W : [0, T] \to L_2(\Omega)$ be a component of the Wiener-Lévy N vector $\boldsymbol{W}$ with

$$\mathrm{E}W^2(t) = \int_0^t b(u)\, du \,, \tag{3.63}$$

where b is real-valued, continuous and nonnegative on $[0, T]$. For W we define the sequence $\{W^{(n)}\}_{n\in\mathbb{N}}$ as

$$W^{(n)}(t) = \int_0^T q_n(t-s)\, dW(s) \,, \quad t \in [0, T] \,, \tag{3.64}$$

which may be seen as a m.s. integral as well as a sample integral. Integration by parts gives

$$W^{(n)}(t) = q_n(t - T)\, W(T) + \int_0^T W(s)\, p_n(t-s)\, ds \,. \tag{3.65}$$

We remark that $W^{(n)}(t)$ behaves qualitatively as

$$\int_0^T W(s)\, p_n(t-s)\, ds$$

but, for computational convenience, the definition (3.64) of $W^{(n)}(t)$ is preferred.

Some of the properties of the sequence $\{W^{(n)}\}_{n\in\mathbb{N}}$ are given below.

1) The sample functions of $W^{(n)}$ are infinitely often differentiable on $[0, T]$ with probability one. This follows from (3.65) and relevant theorems in ordinary real analysis.

2) Sequence $W^{(n)}$ is also smooth in m.s. on $[0, T]$. To verify this, we shall show only that its first m.s. derivative exists. We have, following the mean value theorem and (2.108),

$$\left\| \frac{1}{h} [W^{(n)}(t+h) - W^{(n)}(t)] - \int_0^T p_n(t-s)\, dW(s) \right\|^2$$

$$= \left\| \int_0^T \left\{ \frac{1}{h} [q_n(t+h-s) - q_n(t-s)] - p_n(t-s) \right\} dW(s) \right\|^2$$

$$= \left\| \int_0^T [p_n(t + \vartheta(s) h - s) - p_n(t-s)]\, dW(s) \right\|^2$$

$$= \int_0^T [p_n(t + \vartheta(s) h - s) - p_n(t-s)]^2 b(s)\, ds\,, \qquad |\vartheta(s)| < 1\,.$$

Due to uniform continuity of p_n, the right-hand side tends to zero as $h \to 0$, showing that

$$\frac{1}{h} \{W^{(n)}(t+h) - W^{(n)}(t)\}$$

converges in m.s. as $h \to 0$. Thus $W^{(n)}$ is m.s. differentiable. The existence of higher-order m.s. derivatives can be shown analogously.

3) Following from Theorem 2.20 as applied to (3.64, 65), $\mathrm{E}W^{(n)}(t) = 0$, $t \in [0, T]$.

4) From considerations advanced in Sect. 2.7 $W^{(n)}(t)$, $t \in [0, T]$, is normally distributed.

5) If $t_1 \leqslant t_2 < t_3 \leqslant t_4$, the increments of $W^{(n)}$ on $[t_1, t_2)$ and $[t_3, t_4)$ are orthogonal (and independent) for sufficiently large n. Again following the mean value theorem and (2.108), we have

$$\mathrm{E}\{[W^{(n)}(t_4) - W^{(n)}(t_3)][W^{(n)}(t_2) - W^{(n)}(t_1)]\}$$

$$= \mathrm{E}\left\{ \int_0^T [q_n(t_4 - s) - q_n(t_3 - s)]\, dW(s) \int_0^T [q_n(t_2 - s) - q_n(t_1 - s)]\, dW(s) \right\}$$

$$= \int_0^T [q_n(t_4 - s) - q_n(t_3 - s)][q_n(t_2 - s)\, q_n(t_1 - s)]\, b(s)\, ds$$

$$= (t_4 - t_3)(t_2 - t_1) \int_0^T p_n(\vartheta(s) - s)\, p_n(\vartheta^*(s) - s)\, b(s)\, ds\,,$$

in which $\vartheta(s) \in (t_3, t_4)$ and $\vartheta^*(s) \in (t_1, t_2)$ for all s. Hence, for all s,

$$\vartheta(s) - \vartheta^*(s) > t_3 - t_2 > 0$$

and, if n is sufficiently large, the supports of $p_n(\vartheta(s) - s)$ and $p_n(\vartheta^*(s) - s)$ are disjoint, yielding zero for the expression considered above.

6) Let Ω be the point set of the probability space $\{\Omega, \mathcal{A}, P\}$ on which the random variables of the Wiener-Lévy process $W : [0, T] \to L_2(\Omega)$ are

defined. Using the notations $W(t, \omega)$ and $W^{(n)}(t, \omega)$ in place of $W(t)$ and $W^{(n)}(t)$ as introduced in Sect. 1.4, the following property arises.

At almost all $\omega \in \Omega$, $W^{(n)}(t, \omega) \to W(t, \omega)$ as $n \to \infty$, $t \in [0, T]$, i.e., at any $t \in [0, T]$, the samples of the approximating processes $W^{(n)}$ tend a.s. to the corresponding samples of W. As the samples of W are a.s. not of bounded variation on $[0, T]$, this is seen from (3.64) and ordinary calculus.

7) As $n \to \infty$, $W^{(n)}(t) \to W(t)$ in m.s. uniformly in $t \in [0, T]$. To show this, we first note that

$$\int_0^T h(t-s)\, dW(s) = \int_0^t dW(s) = W(t) .$$

Hence, if $b = \max\limits_{t \in [0,T]} |b(t)|$, (3.62) together with the convergence theorem of Lebesgue give

$$\begin{aligned} \|W^{(n)}(t) - W(t)\|^2 &= \left\| \int_0^T [q_n(t-s) - h(t-s)]\, dW(s) \right\|^2 \\ &= \int_0^T [q_n(t-s) - h(t-s)]^2 b(s)\, ds \\ &\leqslant b \int_{-T}^T [q_n(x) - h(x)]^2 dx \to 0 \end{aligned}$$

as $n \to \infty$, independently of the value of t in $[0, T]$.

8) Property (7) leads to

$$\mathrm{E}\{W^{(n)}(s)\, W^{(n)}(t)\} \to \mathrm{E}\{W(s)\, W(t)\} \qquad \left[= \int_0^s b(u)\, du \quad \text{if} \quad 0 \leqslant s \leqslant t \leqslant T \right].$$

Hence, if $t_i \in [0, T]$, $i = 1, \ldots, k$, the elements of the $k \times k$ matrix $\mathrm{E}\{W^{(n)}(t_i)\, W^{(n)}(t_j)\}$ tend to the corresponding elements of $\mathrm{E}\{W(t_i)\, W(t_j)\}$, $i, j = 1, \ldots. k$, as $n \to \infty$. Also, the characteristics function of the Gaussian joint distribution function of the system $\{W^{(n)}(t_1), \ldots, W^{(n)}(t_k)\}$ tends to that of the system $\{W(t_1), \ldots, W(t_k)\}$ as $n \to \infty$ [see (1.50)].

In summary, properties (1–5) describe $W^{(n)}$ as a realistic mathematical model for a particle executing Brownian motion, whereas properties (6–8) show that $W^{(n)}$ converges to W in a satisfactory manner. Analogously, the components of the Wiener-Lévy N vector may be approximated by Gaussian N vectors with smooth components.

Let us now turn our attention to system (3.3), and consider first the system

$$\begin{aligned} &\frac{d}{dt} \boldsymbol{X}^{(n)}(t) = A(t) \boldsymbol{X}^{(n)}(t) + \frac{d}{dt} \boldsymbol{W}^{(n)}(t) , \quad t \in [0, T] \\ &\boldsymbol{X}^{(n)}(0) = \boldsymbol{C} \end{aligned} \tag{3.66}$$

or, equivalently,

$$X^{(n)}(t) = C + \int_0^t A(s) X^{(n)}(s)\, ds + W^{(n)}(t), \quad t \in [0, T], \tag{3.67}$$

where the N vector $W^{(n)}(t)$ is as defined in (3.64) and the matrix $A(t)$ and the N vector C are the same as in (3.3 or 51).

We recall that the solution to (3.51) may be represented by (3.52). Analogously, the solution to (3.66, 67) may be represented by

$$X^{(n)}(t) = \Phi(t) \left[C + \int_0^t \Phi^{-1}(s)\, dW^{(n)}(s) \right], \quad t \in [0, T]. \tag{3.68}$$

Upon integration by parts, (3.52, 68) may be rewritten as, respectively,

$$X(t) = \Phi(t) \left[C + \int_0^t \Phi^{-1}(s) A(s) W(s)\, ds \right] + W(t), \quad t \in [0, T] \tag{3.69}$$

and

$$X^{(n)}(t) = \Phi(t) \left[C + \int_0^t \Phi^{-1}(s) A(s) W^{(n)}(s)\, ds \right] + W^{(n)}(t),$$

$$t \in [0, T], \tag{3.70}$$

since

$$\Phi^{-1}(t)\Phi(t) = I \quad \text{and} \quad \frac{d}{dt}\Phi(t) = A(t)\Phi(t)$$

yield

$$\left[\frac{d}{dt}\Phi^{-1}(t) \right] \Phi(t) + \Phi^{-1}(t) A(t) \Phi(t) = 0 \quad \text{and}$$

$$\frac{d}{dt}\Phi^{-1}(t) = -\Phi^{-1}(t) A(t).$$

Clearly, (3.69, 70) represent the solutions in the m.s. as well as in the sample sense. Their properties and their interrelationship can be inferred from the properties established earlier for $W^{(n)}(t)$ and $W(t)$. We note especially the following.

a) It follows from (3.70) that the samples of the components of $X^{(n)}$ are smooth and that the components of $X^{(n)}$ are smooth in m.s. on $[0, T]$. Furthermore,

$$\mathrm{E}X^{(n)}(t) = 0, \quad t \in [0, T]$$

and $X^{(n)}(t)$, $t \in [0, T]$, is normally distributed (Sect. 2.7).

b) Equations (3.69, 70) show that

$$X(t) - X^{(n)}(t) = \int_0^t \Phi(t)\Phi^{-1}(s) A(s)[W(s) - W^{(n)}(s)]\, ds$$

$$+ [W(t) - W^{(n)}(t)].$$

Properties (6, 7) concerning $\boldsymbol{W}^{(n)}(t)$ and $\boldsymbol{W}(t)$ yield

$$X_i(t) - X_i^{(n)}(t) \to 0 \quad \text{as} \quad n \to \infty \tag{3.71}$$

a.s. in the sample sense as well as uniformly in t on $[0, T]$ in m.s., $i = 1, \ldots, N$.

It also follows from the arguments similar to those in Property (8) above that the characteristic function of the joint distribution function of the system $\{X_1^{(n)}(t_{1,1}), \ldots, X_1^{(n)}(t_{1,k_1}), \ldots, X_N^{(n)}(t_{N,1}), \ldots, X_N^{(n)}(t_{N,k_N})\}$ tends to that of $\{X_1(t_{1,1}), \ldots, X_1(t_{1,k_1}), \ldots, X_N(t_{N,1}), \ldots, X_N(t_{N,k_N})\}$ as $n \to \infty$.

These results are a special case of a theorem by *Wong* and *Zakai* [3.3]. We see from (a) that $\boldsymbol{X}^{(n)}$ is a realistic mathematical model of a state vector of a physical system, whereas the arguments in (b) show the manner in which $\boldsymbol{X}^{(n)}$ approximates $\boldsymbol{X}$, the solution of (3.3).

On the basis of these discussions, it is seen that using (3.3) as the mathematical model of the basic system in Kalman-Bucy filtering is sufficiently justified. Because of computational advantages as indicated earlier, this idealized system will be used throughout our discourse.

4. The Kalman-Bucy Filter

Equation (3.1 or 3) mathematically describes the dynamic model in the Kalman-Bucy filter, whose output is a stochastic vector process to be estimated on the basis of noise-contaminated observations. To complete the Kalman-Bucy model for the estimation problem, we now define the observation process.

The *observations* $\boldsymbol{Z}(s)$ of $\boldsymbol{X}(s)$ are assumed to be of the type

$$\boldsymbol{Z}(s) = \int_0^s H(u)\boldsymbol{X}(u)\,du + \boldsymbol{W}^*(s) , \quad 0 \leqslant s \leqslant t \leqslant T , \tag{4.1}$$

where H is an $M \times N$ matrix with continuous components and $\boldsymbol{W}^*(s)$ is a Wiener-Lévy M vector, stochastically independent of the Wiener-Lévy process $\boldsymbol{W}(t)$, $t \in [0, T]$, and the initial condition $\boldsymbol{C}$ associated with the dynamic system defined by (3.1). Moreover, $\boldsymbol{W}^*(s)$ is required to satisfy an additional condition, discussed below, which is of central importance to the success of resulting computational algorithms.

The stochastic M vector process

$$\boldsymbol{Z} : [0, t] \rightarrow L_2^M(\Omega)$$

is called the *observation process*. The generalized derivative of $\boldsymbol{W}^*(s)$ in (4.1) is called the *observation noise*.

In Kalman-Bucy filtering, the basic aim is to construct a stochastic process $\hat{\boldsymbol{X}}(t)$, $t \in [0, T]$, defined on the observations $\boldsymbol{Z}(s)$, $0 \leqslant s \leqslant t$, whose components $\hat{X}_i(t)$, $i = 1, \ldots, N$, approximate the components $X_i(t)$ of the output process $\boldsymbol{X}(t)$ in some optimal sense. Finally, in the resulting computational algorithm for $\hat{\boldsymbol{X}}(t)$, the random vectors $\boldsymbol{Z}(s)$ are replaced by the observed data $\boldsymbol{Z}(s, \omega)$, giving the "optimal" reconstruction of $\boldsymbol{X}(t, \omega)$. It is noted that the Kalman-Bucy filter is not given by an explicit prescription of the optimal filter, but as an algorithm suitable for direct evaluation by digital or analog computers.

The filtering procedure is called *discrete* when $\boldsymbol{X}(t)$ is observed at a finite number of time instants. The discrete version of the Kalman-Bucy filter is called the Kalman filter, after Kalman, who was the first to contribute substantially to the topic of discrete nonstationary filtering [4.1].

4.1 Some Preliminaries

Before proceeding to the mathematical development of Kalman-Bucy filtering, we first comment on some properties of the output and observation processes and the underlying estimation procedure.

Consider the output process $\boldsymbol{X}(t)$ as defined by (3.1 or 3). Its mean or the deterministic part is given by (3.42) and can be computed. Now, the deterministic part of the observation process $\boldsymbol{Z}(s)$ is

$$\mathrm{E}\boldsymbol{Z}(s) = \int_0^s H(u)\,\mathrm{E}\boldsymbol{X}(u)\,du \tag{4.2}$$

and the zero-mean random part of $\boldsymbol{Z}(s)$ is thus given by

$$\boldsymbol{Z}_o(s) = \boldsymbol{Z}(s) - \mathrm{E}\boldsymbol{Z}(s) = \int_0^s H(u)\boldsymbol{X}_o(u)\,du + \boldsymbol{W}^*(s)\,, \tag{4.3}$$

where, we recall, $\boldsymbol{X}_o(u) = \boldsymbol{X}(u) - \mathrm{E}\boldsymbol{X}(u)$. We shall reduce the filtering procedure to the approximation of components $X_{oi}(t)$ of $\boldsymbol{X}_o(t)$ by means of functions of the components $Z_{oj}(s)$, $0 \leqslant s \leqslant t$; $j = 1,\ldots,M$, of $\boldsymbol{Z}_o(s)$ (cf. Exercise 4.1).

Additionally, computational considerations suggest that the "best" approximation $\hat{X}_i(t)$ of $X_{oi}(t)$ is based on *linear least-squares estimation*. This means that if $H(\boldsymbol{Z}_o, t)$ is the closure in $L_2(\Omega)$ of the linear hull of the set

$$\{Z_{oj}(s); j = 1,\ldots,M; s \in [0, T]\}\,, \tag{4.4}$$

then $\hat{X}_i(t)$ is the element of $H(\boldsymbol{Z}_o, t)$ satisfying

$$\mathrm{E}\{X_{oi}(t) - \hat{X}_i(t)\}^2 = \min_{Y \in H(\boldsymbol{Z}_o, t)} \mathrm{E}\{X_{oi}(t) - Y\}^2\,. \tag{4.5}$$

According to the projection theorem of Hilbert spaces, this element $\hat{X}_i(t)$ exists, is unique and is the orthogonal projection of $X_i(t)$ onto $H(\boldsymbol{Z}_o, t)$.

We shall show that the elements of $H(\boldsymbol{Z}_o, t)$, the element $\hat{X}_i(t)$ in particular, may be represented as

$$\int_0^t \boldsymbol{f}_i^T(s)\,d\boldsymbol{Z}_o(s)\,, \tag{4.6}$$

where the components of $\boldsymbol{f}_i$ are elements of $L_2([0, t])$, and where the integral is of the type defined in Sect. 4.3.

As $\mathrm{E}\boldsymbol{Z}_o(s) = \boldsymbol{0}$ entails $\mathrm{E}Y = 0$ if $Y \in H(Z_o, t)$, and hence $\mathrm{E}\hat{X}_i(t) = 0$, we also have

$$\mathrm{E}\{X_{oi}(t) - \hat{X}_i(t)\} = 0\,. \tag{4.7}$$

The estimator $\hat{X}_i(t)$ is called the *linear minimum variance estimator* or *Kalman-Bucy estimator* of $X_{oi}(t)$.

By Chebychev's inequality, see (1.68), or by (1.34), as all involved random variables are Gaussian, we may interpret (4.5) so that with resonable probability the difference

$$|X_{oi}(t, \omega) - \hat{X}_i(t, \omega)|$$

of the realizations is also minimal.

We remark without proof that since all random variables involved in the Kalman-Bucy filter belong to a Gaussian system, the function $\hat{X}_i(t)$ also minimizes (4.5) if $Y_i(t)$ stands for arbitrary measurable functions of the random variables $Z_{oj}(s)$.

Finally, the reconstruction of $X_i(t, \omega)$ is

$$\mathrm{E}X_i(t) + \hat{X}_i(t, \omega) ,$$

where $\hat{X}_i(t, \omega)$ is obtained from $\hat{X}_i(t)$ by replacing $\mathbf{Z}_o(s)$ by

$$\mathbf{Z}_o(s, \omega) = \mathbf{Z}(s, \omega) - \mathrm{E}\mathbf{Z}(s) .$$

Here $\mathrm{E}\mathbf{Z}(s)$ is computed according to (4.2) and $\mathbf{Z}(s, \omega)$ is measured.

The central problem, to be treated in the following sections, is to derive an efficient computational algorithm for the linear minimum-variance estimator $\hat{X}_i(t)$ of $X_{oi}(t)$, $i = 1, \ldots, N$; $t \in [0, T]$. In what follows, a change of notation shall be made by writing $X_{oi}(t)$ as $X_i(t)$, assuming that $\boldsymbol{X}(t)$ satisfies (3.51) in accordance with the meaning of the symbols used in Theorem 3.5.

In closing, let us remark that since the Kalman-Bucy filter is given as an algorithm for machine computation and thus a discrete procedure, the question naturally arises as to why not work discretely at the outset by using Kalman's discrete filter. The answer is that in the continuous case the existence and uniqueness of $\hat{X}_i(t, \omega)$ has to be shown first, just as in the theory of differential equations where numerical work is meaningful only if existence and uniqueness of their solutions are assured.

4.1.1 Supplementary Exercise

Exercise 4.1. Let $Z_o : [0, t] \to L_2(\Omega)$ have the properties

$$\mathrm{E}Z_o(s) = 0 \quad \text{at all} \quad s \in [0, t] \quad \text{and}$$

$$Z_o(0) = 0 \quad \text{with probability one.}$$

Let $D \in \mathbb{R}$, $D \neq 0$, and let D also denote the (degenerate) random variable equal to D with probability one. Let $Z : [0, t] \to L_2(\Omega)$ be defined by

$$Z(s) = D + Z_o(s) , \quad s \in [0, t] \quad \text{so} \quad Z(0) = D .$$

Let $[D]$ be the subspace of $L_2(\Omega)$, generated by the degenerate random variable D. It is the set of all multiples of D (and it is one-dimensional). Let

$$[Z_o], [Z] \quad \text{and} \quad [D, Z_o]$$

be the closed linear hulls in $L_2(\Omega)$ of the sets

$$\{Z_o(s); s \in [0, t]\}, \{Z(s); s \in [0, t]\} \quad \text{and} \quad \{D, Z_o(s); s \in [0, t]\},$$

respectively.

Since $D \in [Z]$, we have

$$[Z] = [D, Z_o] = [D] + [Z_o]$$

and since $\mathrm{E}\{DZ_o(s)\} = D\mathrm{E}Z_o(s) = 0$ at all $s \in [0, t]$, we have

$$[D] \perp [Z_o]$$

and hence

$$[Z] = [D] \oplus [Z_o].$$

Let $\mathcal{P}_D$, $\mathcal{P}_{Z_o}$ and $\mathcal{P}_Z$ be the ortogonal projection operators of $L_2(\Omega)$ onto $[D]$, $[Z_o]$ and $[Z]$, respectively. Prove then that for any $X \in L_2(\Omega)$ with $X = \mathrm{E}X + X_o$, we have

$$\mathcal{P}_Z X = \mathrm{E}X + \mathcal{P}_{Z_o} X_o.$$

4.2 Some Aspects of $L_2([a, b])$

We discuss in this section some elements of the theory of space $L_2([a, b])$ which are needed for establishing (4.6). Discussion of some new types of stochastic integrals is needed as well, which will be presented in the next section.

We recall that $L_2([a, b])$ is a set of measurable functions $f: [a, b] \to \mathbb{R}$ such that

$$\int_a^b f^2(t)\, dt < \infty, \tag{4.8}$$

where the integral is a Lebesgue integral. Analogously to Sect. 1.4.1, it can be shown that $L_2([a, b])$ is a vector space under the usual addition of functions and multiplication with scalars.

From

$$|2f(t)\,g(t)| \leqslant f^2(t) + g^2(t) \tag{4.9}$$

it follows that

$$\int_a^b f(t)\,g(t)\,dt \tag{4.10}$$

exists and is finite if f and g are elements of $L_2([a,\,b])$, and it is an inner product in $L_2([a,\,b])$ if almost everywhere (a.e.) identical functions are identified (Sect. 1.4.1).

Cauchy's inequality gives

$$\left[\int_a^b f(t)\,g(t)\,dt\right]^2 \leqslant \int_a^b f^2(t)\,dt \int_a^b g^2(t)\,dt\,. \tag{4.11}$$

In particular, if $g(t) = 1$ a.e., (4.11) becomes

$$\left|\int_a^b f(t)\,dt\right| \leqslant \left[(b-a)\int_a^b f^2(t)\,dt\right]^{1/2}. \tag{4.12}$$

The inner product (4.10) leads to a norm on $L_2([a,\,b])$ provided again that a.e. identical functions are identified. We write

$$\|f\|^2 = \int_a^b f^2(t)\,dt\,. \tag{4.13}$$

To avoid possible confusion in using the same notation for the norms in $L_2([a,\,b])$ and in $L_2(\Omega)$, the norms are sometimes qualified with a subscript by writing $\|X\|_\Omega$ for the norm of X in $L_2(\Omega)$ and $\|f\|_{[a,\,b]}$ for the norm of f in $L_2([a,\,b])$. Using (4.13), (4.11, 12) can now be written respectively in the forms

$$\left|\int_a^b f(t)\,g(t)\,dt\right| \leqslant \|f\|\,\|g\| \quad \text{and} \tag{4.14}$$

$$\left|\int_a^b f(t)\,dt\right| \leqslant \sqrt{b-a}\,\|f\|\,. \tag{4.15}$$

As in Sect. 1.4.1, a distance d may be defined in $L_2([a,\,b])$ by means of this norm as

$$d(f,\,g) = \|f-g\|, \quad f,\,g \text{ in } L_2([a,\,b])\,. \tag{4.16}$$

Also, the triangle inequality gives

$$\|f-h\| \leqslant \|f-g\| + \|g-h\|, \quad f,g,h \text{ in } L_2([a,\,b])\,. \tag{4.17}$$

A sequence $\{f_n\}_{n\in N}, f_n \in L_2([a,\,b])$, is a Cauchy sequence in $L_2([a,\,b])$ if

$$\|f_m - f_n\| \to 0 \tag{4.18}$$

as $m, n \to \infty$. It can be shwon that every Cauchy sequence in $L_2([a, b])$ has a limit in $L_2([a, b])$. Hence, $L_2([a, b])$ is a Hilbert space.

The space $L_2([a, b])$ will always be understood to possess the distance function defined above. Hence, if a sequence $\{f_n\}_{n \in N}$ is given as a Cauchy sequence in $L_2([a, b])$, (4.18) applies. And $f_n \to f$ in $L_2([a, b])$ as $n \to \infty$ means

$$\|f - f_n\| \to 0 \tag{4.19}$$

as $n \to \infty$ or, equivalently,

$$\int_a^b \{f(t) - f_n(t)\}^2 dt \to 0 \quad \text{as} \quad n \to \infty . \tag{4.20}$$

Let $[a', b'] \subset [a, b]$. The indicator function $i_{[a', b']} : [a, b] \to R$ is defined by

$$i_{[a', b']}(t) = \begin{cases} 1 & \text{if} \quad t \in [a', b'] \\ 0 & \text{if} \quad t \notin [a', b'] \end{cases} . \tag{4.21}$$

Let $I([a, b])$ be the linear hull of the set of indicator functions of all subintervals of $[a, b]$. Then, if $f \in I([a, b])$, subdivision points $t_o, \ldots, t_k$ exist with

$$a = t_o < t_1 < \ldots < t_k = b$$

and real numbers $f_1, \ldots, f_k$, such that

$$f(t) = f_j$$

if $t \in [t_{j-1}, t_j], j = 1, \ldots, k$. The elements of $I([a, b])$ are called step functions on intervals. Clearly,

$$I([a, b]) \subset L_2([a, b]) . \tag{4.22}$$

The distance function of $L_2([a, b])$ is induced in $I([a, b])$. Hence, a sequence $\{f_n\}_{n \in N}, f_n \in I([a, b])$ is a Cauchy sequence in $I([a, b])$ if (4.18) holds.

Since Cauchy sequences in $I([a, b])$ in general do not have a limit, it is not a complete space. However, since a Cauchy sequence in $I([a, b])$ is also a Cauchy sequence in $L_2([a, b])$, it has a limit in the complete space $L_2([a, b])$. Furthermore, we have the following.

Theorem 4.1. The linear hull $I([a, b])$ is dense in $L_2([a, b])$.

Its proof may be found, for example, in [Ref. 4.2, p. 198]. Hence, as a result of Theorem 4.1, for any $f \in L_2([a, b])$ there are sequences $\{f_n\}_{n \in N}$, $f_n \in I([a, b])$ which converge to f in $L_2([a, b])$, i.e.,

$$\|f - f_n\| \to 0 \quad \text{as} \quad n \to \infty .$$

4.2.1 Supplementary Exercise

Exercise 4.2. Show that, with the aid of Theorem 4.1, $L_2([a, b])$ is separable, i.e., there is a countable subset of $L_2([a, b])$ which is dense in $L_2([a, b])$.

4.3 Mean-Square Integrals Continued

We consider in this section some new types of m.s. integrals which will be encountered in this chapter. It is assumed that $[a, b] \subset [0, T] \subset \mathbb{R}$ throughout this section. In situations where confusion may arise, inner products and norms in $L_2(\Omega)$ and in $L_2([a, b])$ are denoted by $(\cdot,\cdot)_\Omega$ and $\|\cdot\|_\Omega$ and by $(\cdot,\cdot)_{[a,b]}$ and $\|\cdot\|_{[a,b]}$, respectively.

Theorem 4.2. If $f \in I([a, b])$ and if $X : [a, b] \to L_2(\Omega)$ is m.s. continuous, then

$$\int_a^b f(t)\, dX(t)$$

exists as a m.s. Riemann-Stieltjes integral and, in particular, so does

$$\int_a^b f(t)\, dW(t)\,,$$

where $W : [0, T] \to L_2(\Omega)$ is the Wiener-Lévy process.

Proof. Since f is of bounded variation on $[a, b]$, this statement is contained in Theorem 2.22. □

Theorem 4.3. If $f \in I([a, b])$ and if $X : [a, b] \to L_2(\Omega)$ is continuously differentiable in m.s. then

$$\int_a^b f(t)\, dX(t) = \int_a^b f(t) X'(t)\, dt\,, \tag{4.23}$$

where the first integral is a m.s. Riemann-Stieltjes integral and the second a m.s. Riemann integral.

Proof. Since $f \in I([a, b])$, there are subdivision points $t_o, \ldots, t_k$ of $[a, b]$ such that

$$a = t_o < t_1 < \ldots < t_k = b$$

and real numbers $c_1, \ldots, c_k$ such that

$$f(t) = c_i \quad \text{if} \quad t \in (t_{i-1}, t_i), \qquad i = 1, \ldots, k \, .$$

The integrals in (4.23) exist and it is easily seen that

$$\begin{aligned}\int_a^b f(t)\, dX(t) &= \sum_{i=1}^{k} c_i \{X(t_i) - X(t_{i-1})\} = \sum_{i=1}^{k} c_i \int_{t_{i-1}}^{t_i} X'(t)\, dt \\ &= \sum_{i=1}^{k} \int_{t_{i-1}}^{t_i} f(t) X'(t)\, dt = \int_a^b f(t) X'(t)\, dt \, . \end{aligned}$$

□

Theorem 4.4. If f and g are elements of $I([a, b])$ and if X and Y are m.s. continuously differentiable mappings of $[a, b]$ into $L_2(\Omega)$, then

$$\begin{aligned} &\mathrm{E}\left\{\int_a^b f(s)\, dX(s) \int_a^b g(t)\, dY(t)\right\} \\ &= \iint_{[a,b]^2} f(s) g(t) \mathrm{E}\{X'(s) Y'(t)\}\, ds\, dt \, . \end{aligned} \tag{4.24}$$

Proof. This statement is contained in Theorem 4.3 and in (*d*) of Theorem 2.22, since the latter is also valid if the processes are sectionally m.s. continuous (continuous in m.s. except for a finite number of jump discontinuities). □

Theorem 4.5. If f and g are elements of $I([a, b])$ and if W_i and W_j are components of the Wiener-Lévy N vector with

$$\mathrm{E}\{W_i(s) W_j(t)\} = \int_0^s b_{ij}(u)\, du \, ,$$

b_{ij} being continuous on $[0, T]$ and $0 \leqslant s \leqslant t \leqslant T$, then

$$\mathrm{E}\left\{\int_a^b f(s)\, dW_i(s) \int_a^b g(t)\, dW_j(t)\right\} = \int_a^b f(u) g(u) b_{ij}(u)\, du \, . \tag{4.25}$$

Proof. This result is contained in Theorem 2.35 since it is also valid if the functions involved are sectionally continuous. □

Theorem 4.6. If $\{f_n\}_{n\in\mathbb{N}}$ is a Cauchy sequence in $I([a, b])$, if $X : [a, b] \to L_2(\Omega)$ is m.s. continuously differentiable and if $W : [0, T] \to L_2(\Omega)$ is the Wiener-Lévy process, then

a) $\left\{\int_a^b f_n(s)\, dX(s) = \int_a^b f_n(s) X'(s)\, ds\right\}_{n\in\mathbb{N}}$ and

b) $\left\{\int_a^b f_n(t)\, dW(t)\right\}_{n\in\mathbb{N}}$

are Cauchy sequences in $L_2(\Omega)$.

Proof. Since $|\mathrm{E}\{X'(s)X'(t)\}|$ is continuous on $[a, b]^2$ (Theorem 2.7), it has a maximum M. Thus, according to Theorem 4.4 and on account of (4.15),

$$\begin{aligned}
&\left\| \int_a^b f_m(s) X'(s)\, ds - \int_a^b f_n(s) X'(s)\, ds \right\|_\Omega^2 \\
&= \left\| \int_a^b [f_m(s) - f_n(s)] X'(s)\, ds \right\|_\Omega^2 \\
&= \left| \iint_{[a,b]^2} [f_m(s) - f_n(s)][f_m(t) - f_n(t)] \mathrm{E}\{X'(s)X'(t)\}\, ds\, dt \right| \\
&\leqslant M \iint_{[a,b]^2} |f_m(s) - f_n(s)|\, |f_m(t) - f_n(t)|\, ds\, dt \\
&= M \left\{ \int_a^b |f_m(u) - f_n(u)|\, du \right\}^2 \leqslant M(b-a)\|f_m - f_n\|_{[a,b]}^2 \to 0
\end{aligned}$$

as $m, n \to \infty$. The proof of part (a) is complete.

For part (b), suppose

$$\mathrm{E}\{W(s)W(t)\} = \int_0^s \bar{b}(u)\, du\,, \quad 0 \leqslant s \leqslant t \leqslant T\,,$$

where $\bar{b}$ is continuous on $[0, T]$ and nonnegative. Let B be the maximum of $\bar{b}$ on $[0, T]$. Then, according to Theorem 4.5,

$$\begin{aligned}
&\left\| \int_a^b f_m(s)\, dW(s) - \int_a^b f_n(s)\, dW(s) \right\|_\Omega^2 \\
&= \left\| \int_a^b [f_m(s) - f_n(s)]\, dW(s) \right\|_\Omega^2 \\
&= \int_a^b [f_m(s) - f_n(s)]^2 \bar{b}(s)\, ds \\
&\leqslant B \int_a^b [f_m(s) - f_n(s)]^2\, ds = B\|f_m - f_n\|_{[a,b]}^2 \to 0
\end{aligned}$$

as $m, n \to \infty$. This gives the result (b). □

Definition. If $f \in L_2([a, b])$, if $X : [a, b] \to L_2(\Omega)$ is m.s. continuously differentiable and if $W : [0, T] \to L_2(\Omega)$ is the Wiener-Lévy process, then

$$\int_a^b f(t)\, dX(t) = \lim_{n \to \infty} \int_a^b f_n(t)\, dX(t) \tag{4.26}$$

$$\int_a^b f(t) X'(t)\, dt = \lim_{n \to \infty} \int_a^b f_n(t) X'(t)\, dt \tag{4.27}$$

$$\int_a^b f(t)\, dW(t) = \lim_{n \to \infty} \int_a^b f_n(t)\, dW(t)\,, \tag{4.28}$$

where $\{f_n\}_{n\in N}$ is any Cauchy sequence in $I([a, b])$, converging to f in $L_2([a, b])$.

We remark that Theorem 4.3 has shown that for each $n \in \mathbb{N}$,

$$\int_a^b f_n(t)\, dX(t) = \int_a^b f_n(t) X'(t)\, dt\,.$$

Hence

$$\int_a^b f(t)\, dX(t) = \int_a^b f(t) X'(t)\, dt\,. \tag{4.29}$$

The definition given above is admissible as all limits involved exist by virtue of the foregoing theorems. And if $\{g_n\}_{n\in N}$ is also a Cauchy sequence in $I([a, b])$ converging to f in $L_2([a, b])$, we have

$$\lim_{n\to\infty} \int_a^b f_n(t)\, dX(t) = \lim_{n\to\infty} \int_a^b g_n(t)\, dX(t) \tag{4.30}$$

as, since $f_n - g_n \in I([a, b])$ as well,

$$\left\| \int_a^b f_n(t)\, dX(t) - \int_a^b g_n(t)\, dX(t) \right\|_\Omega^2$$

$$= \left\| \int_a^b [f_n(t) - g_n(t)] X'(t)\, dt \right\|_\Omega^2$$

$$= \left| \iint_{[a,b]^2} [f_n(s) - g_n(s)][f_n(t) - g_n(t)]\, \mathrm{E}\{X'(s) X'(t)\}\, ds\, dt \right|$$

$$\leqslant M \left[\int_a^b |f_n(u) - g_n(u)| du \right]^2 \leqslant M(b-a) \int_a^b [f_n(u) - g_n(u)]^2\, du$$

$$= M(b-a) \|f_n - g_n\|_{[a,b]}^2$$

$$\leqslant M(b-a) \{\|f_n - f\|_{[a,b]} + \|g_n - f\|_{[a,b]}\}^2 \to 0$$

as $n \to \infty$, where M, as before, is the maximum of $|\mathrm{E}\{X'(s) X'(t)\}|$ on $[a, b]^2$. Hence, (4.26, 27) in the above definition are admissible.

Concerning (4.28), if

$$\mathrm{E}\{W(s) W(t)\} = \int_0^s \bar{b}(u)\, du\,, \quad 0 \leqslant s \leqslant t \leqslant T\,,$$

where $\bar{b}$ is nonnegative and continuous on $[a, b]$ with maximum B, the foregoing theorems give

$$\left\| \int_a^b f_n(t)\, dW(t) - \int_a^b g_n(t)\, dW(t) \right\|_\Omega^2$$
$$= \left| \int_a^b [f_n(t) - g_n(t)]^2 \bar{b}(t)\, dt \right|$$
$$\leqslant B \int_a^b [f_n(t) - g_n(t)]^2 dt = B \|f_n - g_n\|_{[a,b]}^2$$
$$\leqslant B[\|f_n - f\|_{[a,b]} + \|g_n - f\|_{[a,b]}]^2 \to 0$$

as $n \to \infty$, thus showing the admissibility of (4.28) in the above definition.

We remark that the integrals in (4.26–28) are not of m.s. Riemann-Stieltjes type, nor are they m.s. Riemann integrals. However, the use of the same notation for the integrals is justified because they are identical to their m.s. counterparts in Chap. 2, in case f is continuous. To show this, let us for the moment introduce the notations

$$\text{RS-}\int_a^b \quad \text{and} \quad \text{L-}\int_a^b$$

for the integrals in m.s. as defined in Chap. 2 and in definitions (4.26–28), respectively.

Theorem 4.7. If $f : [a, b] \to \mathbb{R}$ is continuous, if $X : [a, b] \to L_2(\Omega)$ is m.s. continuously differentiable and if $W : [0, T] \to L_2(\Omega)$ is the Wiener-Lévy process, then

$$\text{a)} \quad \text{RS-}\int_a^b f(t)\, dX(t) = \text{L-}\int_a^b f(t)\, dX(t) \quad \text{and} \tag{4.31}$$

$$\text{b)} \quad \text{RS-}\int_a^b f(t)\, dW(t) = \text{L-}\int_a^b f(t)\, dW(t) \, . \tag{4.32}$$

Proof. Consider the partition p of (2.16). We have the corresponding Riemann-Stieltjes sum

$$S_{f,X}(p) = \sum_{i=1}^k f(t_i')[X(t_i) - X(t_{i-1})] \, .$$

Hence, if $g : [a, b] \to \mathbb{R}$ is defined as

$$g(a) = f(a)$$
$$g(t) = f(t_i') \quad \text{if} \quad t \in (t_{i-1}, t_i]$$

then $g \in I([a, b])$ and

$$S_{f,X}(p) = \text{RS-}\int_a^b g(t)\, dX(t) \, .$$

Now let $\{p_n\}_{n \in \mathbb{N}}$ be a convergent sequence of partitions of $[a, b]$ (so that $\Delta p_n \to 0$ as $n \to \infty$). To each p_n construct the step function $g_n \in I([a, b])$ in the way we constructed g to p. Then, due to uniform continuity of f on $[a, b]$,

$$g_n \to f$$

as $n \to \infty$ a.e., uniformly on $[a, b]$. Thus, we also have

$$(g_n - f)^2 \to 0$$

uniformly on $[a, b]$. Hence,

$$\int_a^b [g_n(t) - f(t)]^2 \, dt \to 0$$

as $n \to \infty$ or

$$g_n \to f$$

in $L_2([a, b])$.

Therefore, through (4.26) and the foregoing theorems, we have

$$S_{f,X}(p_n) = \int_a^b g_n(t) \, dX(t) \to \text{L--}\int_a^b f(t) \, dX(t)$$

in m.s. as $n \to \infty$. As

$$S_{f,X}(p_n) \to \text{RS--}\int_a^b f(t) \, dX(t)$$

also, part (a) of the theorem is proved. Part (b) can be shown analogously. □

Corollary.

$$\text{RS--}\int_a^b f(t) X'(t) \, dt = \text{L--}\int_a^b f(t) X'(t) \, dt \,. \tag{4.33}$$

Proof. Since X is m.s. continuously differentiable on $[a, b]$ and f is continuous, we have, according to Theorem 2.27 and the comments advanced on definitions (4.26–28),

$$\begin{aligned} \text{RS--}\int_a^b f(t) X'(t) \, dt &= \text{RS--}\int_a^b f(t) \, dX(t) \\ &= \text{L--}\int_a^b f(t) \, dX(t) \\ &= \text{L--}\int_a^b f(t) X'(t) \, dt \,. \end{aligned}$$

□

Hence, as we shall work only with RS integrals involving f, X, W satisfying the above conditions, the prefixes "RS–" and "L–" to the integral signs may be omitted without risk of confusion.

We remark that we could have also worked with the above integrals in the foregoing chapters. However, the integrals in Chap. 2 are better adapted to the dynamical systems of Chap. 3, whereas the latter will exactly fit the estimations, as may be seen in the following sections.

The following statement is an immediate result of definitions (4.26–28).

Theorem 4.8. The integrals defined in (4.26–28) have the properties discussed in Sect. 2.4.1, except for partial integration.

Theorem 4.9. If f and g are elements of $L_2([a, b])$, if X and Y are m.s. continuously differentiable mappings of $[a, b]$ into $L_2(\Omega)$ and if W_i and W_j are components of the Wiener-Lévy N vector with

$$\mathrm{E}\{W_i(s)\, W_j(t)\} = \int_0^s b_{ij}(u)\, du\,, \quad 0 \leqslant s \leqslant t \leqslant T\,,$$

where b_{ij} is continuous on $[0, T]$, then

$$\text{a)}\quad \mathrm{E}\left\{\int_a^b f(s)\, dX(s) \int_a^b g(t)\, dY(t)\right\} = \iint_{[a,b]^2} f(s)\, g(t)\, \mathrm{E}\{X'(s)\, Y'(t)\}\, ds\, dt \tag{4.34}$$

and

$$\text{b)}\quad \mathrm{E}\left\{\int_a^b f(s)\, dW_i(s) \int_a^b g(t)\, dW_j(t)\right\} = \int_a^b f(u)\, g(u)\, b_{ij}(u)\, du\,. \tag{4.35}$$

In (4.34, 35), the integrals on the right-hand side are Lebesgue integrals.

Proof. Since $\mathrm{E}\{X'(s)\, Y'(t)\}$ and $b_{ij}(u)$ are continuous on their domains, they have maximum values M and B, respectively. It then also follows that the right-hand sides of (4.34, 35) exist as Lebesgue integrals [see (4.9, 10)].

Now let $\{f_n\}_{n\in N}$ and $\{g_n\}_{n\in N}$ be Cauchy sequences in $I([a, b])$ converging to f and g, respectively, in $L_2([a, b])$. Because of continuity of the inner product and as a result of Theorems 2.22 and 4.3,

$$\begin{aligned}
\mathrm{E}\left\{\int_a^b f(s)\, dX(s) \int_a^b g(t)\, dY(t)\right\} &= \mathrm{E}\left\{\lim_{m\to\infty} \int_a^b f_m(s)\, dX(s) \lim_{n\to\infty} \int_a^b g_n(t)\, dY(t)\right\} \\
&= \lim_{\substack{m\to\infty\\ n\to\infty}} \mathrm{E}\left\{\int_a^b f_m(s)\, X'(s)\, ds \int_a^b g_n(t)\, Y'(t)\, dt\right\} \\
&= \lim_{\substack{m\to\infty\\ n\to\infty}} \iint_{[a,b]^2} f_m(s)\, g_n(t)\, \mathrm{E}\{X'(s)\, Y'(t)\}\, ds\, dt \\
&= \iint_{[a,b]^2} f(s)\, g(t)\, \mathrm{E}\{X'(s)\, Y'(t)\}\, ds\, dt\,.
\end{aligned}$$

The latter equality holds since

$$\begin{aligned}
&\left| \iint_{[a,b]^2} [f(s)g(t) - f_m(s)g_n(t)] \mathrm{E}\{X'(s)Y'(t)\}\, ds\, dt \right| \\
&\quad \leqslant M \iint_{[a,b]^2} |f(s)g(t) - f_m(s)g_n(t)|\, ds\, dt \\
&\quad \leqslant M\Big\{ \iint_{[a,b]^2} |f(s)g(t) - f_m(s)g(t)|\, ds\, dt \\
&\qquad + \iint_{[a,b]^2} |f_m(s)g(t) - f_m(s)g_n(t)|\, ds\, dt\Big\} \\
&\quad = M\left\{ \int_a^b |f(s) - f_m(s)|\, ds \int_a^b |g(t)|\, dt + \int_a^b |f_m(s)|\, ds \int_a^b |g(t) - g_n(t)|\, dt \right\} \\
&\quad \leqslant M\{\sqrt{b-a}\,\|f - f_m\|\sqrt{b-a}\,\|g\| + \sqrt{b-a}\,\|f_m\|\sqrt{b-a}\,\|g - g_n\|\} \\
&\quad \to 0
\end{aligned}$$

as $m, n \to \infty$. In the above, we used (4.15) and the fact that $\|f_m\|$ is bounded as

$$\|f_m\| \leqslant \|f_m - f\| + \|f\| .$$

To show (b), by virtue of continuity of the inner product and due to Theorem 4.5,

$$\begin{aligned}
&\mathrm{E}\left\{ \int_a^b f(s)\, dW_i(s) \int_a^b g(t)\, dW_j(t) \right\} \\
&\quad = \mathrm{E}\left\{ \lim_{m\to\infty} \int_a^b f_m(s)\, dW_i(s) \lim_{n\to\infty} \int_a^b g_n(t)\, dW_j(t) \right\} \\
&\quad = \lim_{\substack{m\to\infty \\ n\to\infty}} \mathrm{E}\left\{ \int_a^b f_m(s)\, dW_i(s) \int_a^b g_n(t)\, dW_j(t) \right\} \\
&\quad = \lim_{\substack{m\to\infty \\ n\to\infty}} \int_a^b f_m(u)\, g_n(u)\, b_{ij}(u)\, du \\
&\quad = \int_a^b f(u)\, g(u)\, b_{ij}(u)\, du .
\end{aligned}$$

The latter equality holds since, using (4.14) and due to boundedness of $\|g_n\|$,

$$\begin{aligned}
&\left| \int_a^b f_m(u)\, g_n(u)\, b_{ij}(u)\, du - \int_a^b f(u)\, g(u)\, b_{ij}(u)\, du \right| \\
&\quad \leqslant B \int_a^b |f_m(u)\, g_n(u) - f(u)\, g(u)|\, du \\
&\quad \leqslant B\left\{ \int_a^b |f_m(u)\, g_n(u) - f(u)\, g_n(u)|\, du + \int_a^b |f(u)\, g_n(u) - f(u)\, g(u)|\, du \right\} \\
&\quad \leqslant B\{\|f_m - f\|\,\|g_n\| + \|f\|\,\|g_n - g\|\} \to 0
\end{aligned}$$

as $m, n \to \infty$. □

We conclude this section with an application of continuity of the inner product, which will be applied to many situations to follow.

Theorem 4.10. If $\{X_m\}_{m\in \mathbb{N}}$ and $\{Y_n\}_{n\in \mathbb{N}}$ are sequences in $L_2(\Omega)$ converging in $L_2(\Omega)$ to X and Y, respectively, and if $\mathrm{E}\{X_m Y_n\} = 0$ for all m and n in $\mathbb{N}$, then

$$\mathrm{E}\{XY\} = 0 . \tag{4.36}$$

Proof. It follows from continuity of the inner product in $L_2(\Omega)$ that

$$\mathrm{E}\{XY\} = \mathrm{E}\left\{\lim_{m\to\infty} X_m \lim_{n\to\infty} Y_n\right\} = \lim_{\substack{m\to\infty\\ n\to\infty}} \mathrm{E}\{X_m Y_n\} = 0 . \qquad \square$$

Example. If $f \in L_2([0, t])$, if $W : [0, t] \to L_2(\Omega)$ is the Wiener-Lévy process and if X is an element of $L_2(\Omega)$, stochastically independent of W on $[0, t]$, then

$$\mathrm{E}\left\{X \int_0^t f(s)\, dW(s)\right\} = 0 \tag{4.37}$$

as seen by applying (4.28) and by writing $\int_0^t f_n(x)\, dW(x)$, $f_n \in I[0, t]$, as a sum of random variables.

4.4 Least-Squares Approximation in Euclidean Space

Before proceeding to the continuous, infinite-dimensional case, let us first recall the following simple problem.

Given the correlations

$$\begin{matrix} \mathrm{E}Z_1^2 & \mathrm{E}\{Z_1Z_2\} \ldots & \mathrm{E}\{Z_1Z_n\} \ldots & \mathrm{E}\{Z_1X\} \\ & \mathrm{E}Z_2^2 \quad \ldots & \mathrm{E}\{Z_2Z_n\} \ldots & \mathrm{E}\{Z_2X\} \\ & \ldots & \ldots & \ldots \ldots \\ & & \mathrm{E}Z_n^2 \quad \ldots & \mathrm{E}\{Z_nX\} \\ & & & \ldots \ldots \end{matrix}$$

of elements $X, Z_1, Z_2, \ldots, Z_n, \ldots$ of $L_2(\Omega)$, find at each $n \in \mathbb{N}$ the real numbers $x_{n1}, x_{n2}, \ldots, x_{nn}$ such that the linear combination (Fig. 4.1)

$$Y_n = \hat{X}(n) = x_{n1} Z_1 + \ldots + x_{nn} Z_n \tag{4.38}$$

minimizes $\mathrm{E}\{X - Y_n\}^2$, $Y_n \in L(Z_1, \ldots, Z_n)$.

The solution to this least-squares problem requires that $\hat{X}(n)$ be the orthogonal projection of X onto the subspace $L(Z_1, \ldots, Z_n)$ of $L_2(\Omega)$

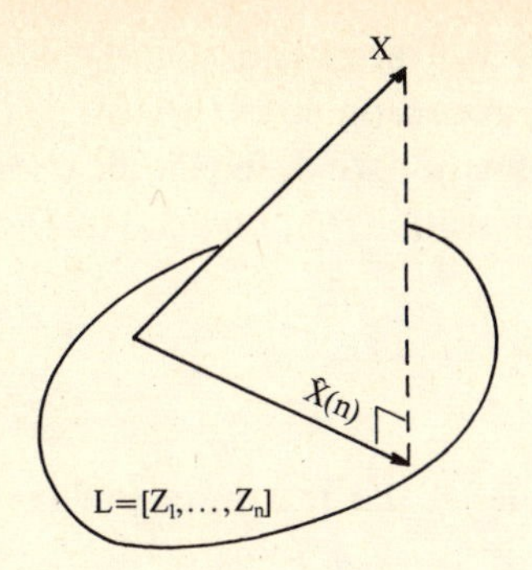

Fig. 4.1. Orthogonal projection $\hat{X}(n)$

spanned by $Z_1, \ldots, Z_n$. The necessary and sufficient condition is

$$X - \hat{X}(n) \perp Z_k \,, \quad k = 1, \ldots, n \,, \quad \hat{x}(n) \in L(Z_1, \ldots, Z_n)$$

That is

$$\mathrm{E}\{[X - \hat{X}(n)]\, Z_k\} = 0 \,, \quad k = 1, \ldots, n, \text{ with (4.38), or}$$

$$\begin{bmatrix} \mathrm{E}Z_1^2 & \mathrm{E}\{Z_1 Z_2\} & \ldots \mathrm{E}\{Z_1 Z_n\} \\ \mathrm{E}\{Z_2 Z_1\} & \mathrm{E}Z_2^2 & \ldots \mathrm{E}\{Z_2 Z_n\} \\ \ldots & \ldots & \ldots \quad \ldots \\ \mathrm{E}\{Z_n Z_1\} & \mathrm{E}\{Z_n Z_2\} & \ldots \quad \mathrm{E}Z_n^2 \end{bmatrix} \begin{bmatrix} x_{n1} \\ x_{n2} \\ \ldots \\ x_{nn} \end{bmatrix} = \begin{bmatrix} E\{Z_1 X\} \\ \mathrm{E}\{Z_2 X\} \\ \ldots \\ \mathrm{E}\{Z_n X\} \end{bmatrix} . \tag{4.39}$$

At each $n \in \mathbb{N}$, system (4.39) always has a solution set $x_{n1}, \ldots, x_{nn}$. It is unique if and only if the set $Z_1, \ldots, Z_n$ is linearly independent. Whether or not this solution is unique, its substitution into (4.38) produces a unique $\hat{X}(n)$.

Computing $\hat{X}(n)$ successively at $n = 1, 2, \ldots$, the solution $x_{n1}, \ldots, x_{nn}$ at n is in general different from the first n elements of the solution $x_{(n+1)1}, \ldots, x_{(n+1)(n+1)}$ at $n + 1$. Thus, it is usually more efficient first to orthogonalize the set $Z_1, Z_2, \ldots$ using, say, the Gram-Schmidt procedure, resulting in $Z_1^*, Z_2^*, \ldots$ such that

$$\mathrm{E}\{Z_m^* Z_n^*\} = \delta_{mn} \,, \quad \delta_{mn} = \begin{cases} 1 & \text{if} \quad m = n \\ 0 & \text{if} \quad m \neq n \end{cases}$$

and

$$L(Z_1, \ldots, Z_n) = L(Z_1^*, \ldots, Z_{n^*}^*) \,, \quad n = 1, 2, \ldots \,,$$

where $n^* \leqslant n$ ($n^* = n$ if $Z_1, \ldots, Z_n$ is a linearly independent set). Now,

$$\hat{X}(n) = (\mathrm{E}\{XZ_1^*\})\, Z_1^* + \ldots + (\mathrm{E}\{XZ_n^*\})\, Z_n^* \quad \text{and} \tag{4.40}$$

$$\hat{X}(n+1) = \hat{X}(n) + (\mathrm{E}\{XZ_{n+1}^*\})\, Z_{n+1}^* \,,$$

showing that the coefficients of $Z_1^*, \ldots, Z_n^*$ in $\hat{X}(n+1)$ are identical to those in $\hat{X}(n)$ and that $\hat{X}(n+1)$ may be computed by recursion on $\hat{X}(n)$.

This is the idea basic to many approximation methods and, in particular, to the discrete Kalman filter and to *Kailath's* "innovations" approach [4.3].

4.4.1 Supplementary Exercises

Exercise 4.3. Verify the following: (a) The rank of the matrix in (4.39) (matrix of Gram) is equal to the dimension of $L(Z_1, \ldots, Z_n)$. In particular, it is invertible if and only if $Z_1, \ldots, Z_n$ is a linearly independent set. (b) The system (4.39) is solvable for any $X \in L_2(\Omega)$ and, whether or not its solution is unique, $\hat{X}(n)$ in (4.38) is unique.

Exercise 4.4. Show that the (Gram) matrix of system (4.39) is diagonalizable and has real and nonnegative eigenvalues. Show that these eigenvalues are positive if and only if the matrix is nonsingular.

4.5 A Representation of Elements of $H(\mathbf{Z}, t)$

As discussed in Sect. 4.1, the Kalman-Bucy estimator of the state vector $\mathbf{X}(t)$ of the dynamical system (3.51) is the N vector $\hat{\mathbf{X}}(t)$, whose components $\hat{X}_i(t)$ are orthogonal projections of the components $X_i(t)$ of $\mathbf{X}(t)$ onto the subspace of $L_2(\Omega)$ generated by the components $Z_j(s)$ of the observation process $\mathbf{Z}(s)$ at all $s \in [0, t]$ and $j = 1, \ldots, m$. This subspace, no longer finite-dimensional, is the closure in $L_2(\Omega)$ of the linear hull $L(\mathbf{Z}, t)$ of the set of elements $Z_j(s)$; it is again a Hilbert space, and it is denoted by $H(\mathbf{Z}, t)$. Now the projection theorem of Hilbert spaces is applicable, i.e., for any $X \in L_2(\Omega)$, there is exactly one element $\hat{X} \in H(\mathbf{Z}, t)$ such that $X - \hat{X}$ is orthogonal to $H(\mathbf{Z}, t)$ and, equivalently, such that

$$\min_{Y \in H(\mathbf{Z}, t)} \|X - Y\|$$

is attained in $Y = \hat{X}$ [4.4, 5].

To enable computations involved in Kalman-Bucy filtering, in this section we shall be concerned with establishing a representation of the elements of $H(\mathbf{Z}, t)$.

The space $H(\mathbf{Z}, t)$ does not consist merely of linear combinations of elements $Z_j(s)$; the limits in $L_2(\Omega)$ of Cauchy sequences of linear combinations of the $Z_j(s)$ also belong to $H(\mathbf{Z}, t)$. Since any linear combination of the

$Z_j(s)$ may also be seen as a m.s. limit of a Cauchy sequence (of identical elements of $L(\mathbf{Z}, t)$, for example), $H(\mathbf{Z}, t)$ is the set of limits in $L_2(\Omega)$ of all Cauchy sequences whose elements belong to $L(\mathbf{Z}, t)$.

We shall proceed in several steps.

A) Let $X : \mathbb{R} \to L_2(\Omega)$ be an arbitrary second-order process. Let $L(X)$ be the linear hull of the set $\{X(t), t \in \mathbb{R}\}$ and let $H(X)$ be the closure of $L(X)$ in $L_2(\Omega)$. Any element of $L(X)$ is then some linear combination

$$x_1 X(t_1) + \ldots + x_k X(t_k) ,$$

where $x_i \in \mathbb{R}$, $t_i \in \mathbb{R}$, $i = 1, \ldots, k$, and each element of $H(X)$ is the m.s. limit of a Cauchy sequence of elements of $L(X)$.

It is not possible to give a useful representation of the elements of $H(X)$ without further information on the process X. Suppose, for example, that X is m.s. differentiable at t_o. Then

$$\left\{\frac{X(t_o + 1/n) - X(t_o)}{1/n}\right\}_{n \in \mathbb{N}}$$

is a Cauchy sequence in $L(X)$ with m.s. limit

$$\left[\frac{d}{dt} X(t)\right]_{t = t_o}$$

belonging to $H(X)$.

As another example, suppose that $\mathrm{E}\{X(s) X(t)\}$ is of bounded variation on $[a, b]^2$ and that $f : [a, b] \to \mathbb{R}$ is continuous. Then, the m.s. integral

$$\int_a^b f(t)\, dX(t)$$

belongs to $H(X)$ as it is the m.s. limit of a Cauchy sequence of Riemann-Stieltjes sums, the latter being elements of $L(X)$.

B) Let $X : [0, t] \to L_2(\Omega)$ be m.s. continuous with $X(0) = 0$ and let $L(X, t)$ be the linear hull of the set $\{X(s), s \in [0, t]\}$. If X_o is an element of $L(X, t)$, then there are numbers $x_1, \ldots, x_k$ in $\mathbb{R}$ and numbers $s_1, \ldots, s_k$ in $[0, t]$ such that

$$X_o = x_1 X_1(s_1) + \ldots + x_k X(s_k) . \tag{4.41}$$

We shall show that there is a step function on intervals $f \in I([0, t])$ such that (Sect. 4.2)

$$X_o = \int_0^t f(s)\, dX(s) . \tag{4.42}$$

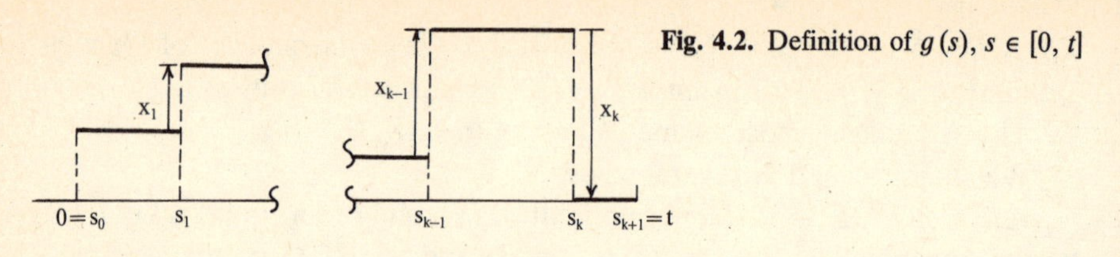

Fig. 4.2. Definition of $g(s)$, $s \in [0, t]$

The simple algebra is as follows. Let $0 = s_o < s_1 < \ldots < s_{k+1} = t$ and define $g : [0, t] \to \mathbb{R}$ recursively as (Fig. 4.2)

$$g(s) = 0 \qquad \text{if } s \in [s_k, s_{k+1}]$$

$$g(s') - g(s) = x_i \qquad \text{if } s \in [s_{i-1}, s_i) \ , \quad s' \in [s_i, s_{i+1}) \ ,$$

$$i = 1, \ldots, k \ .$$

Clearly,

$$\sum_{i=1}^{k} x_i X(s_i) = \int_0^t X(s)\, dg(s) \ , \tag{4.43}$$

where, as shown in Sect. 4.3, the integral may be seen as a m.s. Riemann-Stieltjes integral. Partial integration of (4.43) thus gives

$$\int_0^t X(s)\, dg(s) = [X(s)\, g(s)]_0^t - \int_0^t g(s)\, dX(s)$$

$$= - \int_0^t g(s)\, dX(s)$$

and, letting $f = -g$, we have

$$X_o = \sum_{i=1}^{k} x_i X(s_i) = \int_0^t f(s)\, dX(s) \ .$$

C) Let $W : [0, T] \to L_2(\Omega)$ be the standard Wiener-Lévy process with $EW(s) = 0$ and $E\{W(s)\, W(t)\} = s$, $0 \leqslant s \leqslant t \leqslant T$. Since W is m.s. continuous on $[0, T]$ and $W(0) = 0$, the result of B is applicable, i.e., any element of the linear hull $L(W, t)$ of the set $\{W(s), s \in [0, t]\}$ may be represented as a m.s. Riemann-Stieltjes integral in the form

$$\int_0^t g(s)\, dW(s) \ ,$$

where $g : [0, t] \to \mathbb{R}$ is an element of $I([0, t])$ (Sect. 4.2).

Let $H(W, t)$ be the closure of $L(W, t)$ in $L_2(\Omega)$ and let $X \in H(W, t)$. Then there is a Cauchy sequence $\{X_n\}_{n \in \mathbb{N}}$ in $L(W, t)$ which converges to X in m.s. For any X_n, there is a $f_n \in I([0, t])$ such that

$$X_n = \int_0^t f_n(s)\, dW(s)$$

and thus

$$\left\{ \int_0^t f_n(s)\, dW(s) \right\}_{n \in \mathbb{N}}$$

is a Cauchy sequence in $L(W, t)$. Equivalently, we have, as $m, n \to \infty$,

$$\left\| \int_0^t f_m(s)\, dW(s) - \int_0^t f_n(s)\, dW(s) \right\|_\Omega^2 \to 0$$

$$\Leftrightarrow \left\| \int_0^t [f_m(s) - f_n(s)]\, dW(s) \right\|_\Omega^2 \to 0$$

$$\Leftrightarrow \int_0^t [f_m(s) - f_n(s)]^2\, ds \to 0$$

$$\Leftrightarrow \|f_m - f_n\|_{[a, b]} \to 0 .$$

Hence, $\{X_n\}_{n \in \mathbb{N}}$ is a Cauchy sequence in $L(W, t)$ if and only if $\{f_n\}_{n \in \mathbb{N}}$ is a Cauchy sequence in $I([0, t])$.

Let f be the limit of f_n in the complete space $L_2([0, t])$. Then, according to definition (4.28),

$$X = \lim_{n \to \infty} X_n = \lim_{n \to \infty} \int_0^t f_n(s)\, dW(s) = \int_0^t f(s)\, dW(s) . \tag{4.44}$$

If also $h \in L_2([0, t])$ such that

$$X = \int_0^t h(s)\, dW(s) , \quad \text{then} \tag{4.45}$$

$$\int_0^t f(s)\, dW(s) = \int_0^t h(s)\, dW(s)$$

implies

$$\int_0^t [f(s) - h(s)]\, dW(s) = 0$$

and, it follows from Theorem 4.9 that

$$0 = \mathrm{E} \left\{ \int_0^t [f(s) - h(s)]\, dW(s) \int_0^t [f(s') - h(s')]\, dW(s') \right\}$$

$$= \int_0^t [f(s) - h(s)]^2\, ds .$$

Thus

$$f = h \text{ a.e.}$$

Hence, for each $X \in H(W, t)$, there is a (a.e.) unique $f \in L_2([0, t])$ such that

$$X = \int_0^t f(s)\, dW(s) .$$

Conversely, if $f \in L_2([0, t])$, it is easily seen that

$$\int_0^t f(s)\,dW(s) \in H(w, t) .$$

D) Let $R : [0, T] \to L_2(\Omega)$ be m.s. continuous and define $Y : [0, T] \to L_2(\Omega)$ by

$$\forall t \in [0, T], \qquad Y(t) = \int_0^t R(s)\,ds . \tag{4.46}$$

Then, Y is m.s. continuously differentiable and, as seen from Theorem 2.24,

$$\forall t \in [0, T], \qquad \frac{d}{dt} Y(t) = R(t) . \tag{4.47}$$

Moreover,

$$Y(0) = 0 . \tag{4.48}$$

Let $W : [0, T] \to L_2(\Omega)$ be the standard Wiener-Lévy process with $\mathrm{E}W(t) = 0$ and $\mathrm{E}\{W(s)\,W(t)\} = s$, $0 \leqslant s \leqslant t \leqslant T$. Suppose that W and R are stochastically independent. Then (Sect. 1.3.1)

$$\mathrm{E}\{W(s)\,R(t)\} = \mathrm{E}W(s)\,\mathrm{E}R(t) = 0 , \quad (s, t) \in [0, T]^2 \tag{4.49}$$

and, by Theorem 4.10,

$$\mathrm{E}\{W(s)\,Y(t)\} = 0 , \quad (s, t) \in [0, T]^2 . \tag{4.50}$$

Define $Z : [0, T] \to L_2(\Omega)$ by

$$\forall t \in [0, T], \quad Z(t) = Y(t) + W(t) . \tag{4.51}$$

Then Z is m.s. continuous on $[0, T]$ and $Z(0) = 0$. Let $L(Z, t)$ be the linear hull of the set $\{Z(s),\ s \in [0, t]\}$. The result in B applies. That is, for any element $V \in L(Z, t)$, there is a step function on intervals $g \in I([0, t])$ such that

$$V = \int_0^t g(s)\,dZ(s) , \tag{4.52}$$

where the integral is a m.s. Riemann-Stieltjes integral. And, from the elementary properties discussed in Chap. 2, we have

$$\begin{aligned} V &= \int_0^t g(s)\,dZ(s) = \int_0^t g(s)\,d[Y(s) + W(s)] \\ &= \int_0^t g(s)\,dY(s) + \int_0^t g(s)\,dW(s) \\ &= \int_0^t g(s)\,R(s)\,ds + \int_0^t g(s)\,dW(s) . \end{aligned} \tag{4.53}$$

Let $H(Z, t)$ be the closure of $L(Z, t)$ in $L_2(\Omega)$ and let $X \in H(Z, t)$. Then there is a Cauchy sequence $\{X_n\}_{n \in \mathbb{N}}$ in $L(Z, t)$ converging to X in m.s. For any X_n, there is a $f_n \in I([0, t])$ such that

$$X_n = \int_0^t f_n(s)\, dZ(s) \quad \text{and thus}$$

$$\left\{ \int_0^t f_n(s)\, dZ(s) \right\}_{n \in \mathbb{N}}$$

is a Cauchy sequence in $L(Z, t)$. Basing on these considerations, we can make the following equivalent statements as $m, n \to \infty$:

$$\left\| \int_0^t f_m(s)\, dZ(s) - \int_0^t f_n(s)\, dZ(s) \right\|_\Omega^2 \to 0$$

$$\Leftrightarrow \quad \left\| \int_0^t [f_m(s) - f_n(s)]\, dZ(s) \right\|_\Omega^2 \to 0$$

$$\Leftrightarrow \quad \mathrm{E}\left\{ \int_0^t [f_m(s) - f_n(s)]\, dY(s) + \int_0^t [f_m(s) - f_n(s)]\, dW(s) \right\}^2 \to 0$$

$$\Leftrightarrow \quad \mathrm{E}\left\{ \int_0^t [f_m(s) - f_n(s)]\, dY(s) \right\}^2 + \mathrm{E}\left\{ \int_0^t [f_m(s) - f_n(s)]\, dW(s) \right\}^2 \to 0$$

[as a result of (4.50) and of Theorem 4.10]

$$\Leftrightarrow \quad \mathrm{E}\left\{ \int_0^t [f_m(s) - f_n(s)]\, dY(s) \right\}^2 \to 0 \quad \text{and} \tag{4.54}$$

$$\mathrm{E}\left\{ \int_0^t [f_m(s) - f_n(s)]\, dW(s) \right\}^2 = \int_0^t [f_m(s) - f_n(s)]^2\, ds \to 0 \tag{4.55}$$

since both parts are nonnegative.

Equivalently, we obtain that $\{f_n\}_{n \in \mathbb{N}}$ is a Cauchy sequence in $I([0, t])$ by using results in C. Hence, there is an element $f \in L_2([0, t])$ such that

$$f_n \to f \tag{4.56}$$

in $L_2([0, t])$ as $n \to \infty$.

Conversely, following Theorem 4.6 and definitions (4.26, 28),

$$\int_0^t f_n(s)\, dY(s) \to \int_0^t f(s)\, dY(s) \quad \text{and} \tag{4.57}$$

$$\int_0^t f_n(s)\, dW(s) \to \int_0^t f(s)\, dW(s) \tag{4.58}$$

in m.s. s $n \to \infty$. Hence (also see Theorem 4.8),

$$
\begin{aligned}
X &= \lim_{n\to\infty} X_n = \lim_{n\to\infty} \int_0^t f_n(s)\,dZ(s) \\
&= \lim_{n\to\infty} \int_0^t f_n(s)\,dY(s) + \lim_{n\to\infty} \int_0^t f_n(s)\,dW(s) \\
&= \int_0^t f(s)\,dY(s) + \int_0^t f(s)\,dW(s) \\
&= \int_0^t f(s)\,d[Y(s) + W(s)] \\
&= \int_0^t f(s)\,dZ(s)\,.
\end{aligned} \tag{4.59}
$$

The function $f \in L_2([0, t])$ is uniquely defined by $X \in H(Z, t)$. For, if we also have

$$X = \int_0^t h(s)\,dZ(s)\,, \quad h \in L_2([0, t])\,,$$

we see that, using results in Sect. 4.3,

$$0 = \int_0^t [f(s) - h(s)]\,dZ(s)$$

implies

$$
\begin{aligned}
0 &= \left\| \int_0^t [f(s) - h(s)]\,dZ(s) \right\|^2 \\
&= \mathrm{E}\left\{ \int_0^t [f(s) - h(s)]\,R(s)\,ds + \int_0^t [f(s) - h(s)]\,dW(s) \right\}^2 \\
&= \mathrm{E}\left\{ \int_0^t [f(s) - h(s)]\,R(s)\,ds \right\}^2 + \int_0^t [f(s) - h(s)]^2\,ds
\end{aligned}
$$

which gives, since the first term is nonnegative,

$$\int_0^t [f(s) - h(s)]^2\,ds = 0 \quad \text{or}$$

$$f = h \quad \text{a.e.}$$

It is noteworthy that the results given above follow because of the presence of the Wiener-Lévy process W and its independence of the process R (or Y). For, if there were no noise, we would have only (4.54) or

$$
\begin{aligned}
0 &\leq \mathrm{E}\left\{ \int_0^t [f_m(s) - f_n(s)]\,R(s)\,ds \right\}^2 \\
&= \int_0^t du \int_0^t dv\,[f_m(u) - f_n(u)][f_m(v) - f_n(v)]\,\mathrm{E}\{R(u)\,R(v)\} \to 0
\end{aligned} \tag{4.60}
$$

as $m, n \to \infty$. However, (4.60) does not imply

$$\|f_m(u) - f_n(u)\| \to 0$$

as $m, n \to \infty$ as the following simple example shows.

There is the trivial situation in which $R = 0$. Or, slightly less trivial and suitable for generalizations, define $R : [0, 2\pi] \to L_2(\Omega)$ by

$$\forall t \in [0, 2\pi], \quad R(t) = A \sin t, \quad \text{where}$$

$$A \in L_2(\Omega), \mathrm{E}A^2 = 1 .$$

Now let $\{f_n\}_{n \in \mathbb{N}}$ be any sequence of the functions f_n defined by

$$\forall t \in [0, 2\pi], f_n(t) = a_n$$

with $a_n \in \mathbb{R}$ and $n \in \mathbb{N}$. Then

$$\int_0^{2\pi} [f_m(t) - f_n(t)] R(t)\, dt = (a_m - a_n) A \int_0^{2\pi} \sin t\, dt = 0$$

and thus (4.60) holds, whereas $\{f_n\}_{n \in \mathbb{N}}$ as defined above may not be any Cauchy sequence in $I([0, 2\pi])$.

Summarizing, we have shown that for each $X \in H(Z, t)$, there is a unique $f \in L_2([0, t])$ such that

$$X = \int_0^t f(s)\, dZ(s) \tag{4.61}$$

due to the presence of W in Z. And it is easily seen that for any $f \in L_2([0, t])$, (4.61) belongs to $H(Z, t)$.

E) Finally, let $\mathbf{Z} : [0, t] \to L_2^M(\Omega)$, $t \in [0, T]$, be the M vector observation process in the Kalman-Bucy filter as given by (4.1), i.e.,

$$\mathbf{Z}(s) = \int_0^s H(u)\mathbf{X}(u)\, du + \mathbf{W}^*(s), \quad s \in [0, t], \tag{4.62}$$

where H is an $M \times N$ matrix whose elements are continuous real functions on $[0, t]$, $\mathbf{X} : [0, t] \to L_2^N(\Omega)$ is the state N vector of the dynamical system (3.51) with m.s. continuous components and $\mathbf{W}^* : [0, t] \to L_2^M(\Omega)$ is a Wiener-Lévy M vector process such that $W_j^*(u)$ and $X_i(s)$ are stochastically independent for all i, j, u and s in their respective domains.

We recall that

$$\mathrm{E}\{\mathbf{W}^*(s_1)\mathbf{W}^{*T}(s_2)\} = \int_0^s B^*(u)\, du, \quad s = \min(s_1, s_2), \tag{4.63}$$

where the $M \times M$ matrix $B^*(u)$ has continuous real components, is symmetric and $B^*(u) \geq 0$ at any $u \in [0, t]$ (Sect. 2.6.1).

In (D) we stressed the importance of the noise process in our considerations. Here again, the presence of W^* in the observation process is of crucial importance. In what follows, we further assume

$$B^*(u) > 0\,, \quad u \in [0, t] \tag{4.64}$$

in the sense that

$$x^T B^*(u) x > 0\,, \quad x \in \mathbb{R}^M\,, \quad x \neq 0\,, \quad u \in [0, t]\,. \tag{4.65}$$

This assumption does not limit its applicability in practice for, continuing the comments following the statement of Theorem 2.34, if

$$\det B^*(u) = 0\,,$$

then the number of components of the observation process must be reduced so that the components of the remaining column vector "$Q(u)\,dW^*(u)$" are linearly independent as functions of $\omega \in \Omega$.

Before proceeding, let us first prove the following.

Lemma. If $B^*(u)$ satisfies condition (4.64), there are real numbers $\underline{\lambda}$ and $\overline{\lambda}$, $0 < \underline{\lambda} \leqslant \overline{\lambda}$, such that

$$\underline{\lambda} x^T x \leqslant x^T B^*(u) x \leqslant \overline{\lambda} x^T x\,, \quad x \in \mathbb{R}^M\,, \quad u \in [0, t]\,. \tag{4.66}$$

Proof. Let $S = \{x | x \in \mathbb{R}^M,\, x^T x = 1\}$ and $C = S \times [0, t]$. Then, according to a theorem of Tychonoff [4.4], C is a compact set with respect to the product topology in $S \times [0, t]$. The mapping $F: C \to \mathbb{R}$ defined by

$$F(x, u) = x^T B^*(u) x$$

is continuous (with respect to the product topology) and positive. Hence, F has a minimum $\underline{\lambda} > 0$ and a maximum $\overline{\lambda}$ on the compact set C, i.e.,

$$0 < \underline{\lambda} \leqslant x^T B^*(u) x \leqslant \overline{\lambda}\,. \tag{4.67}$$

Since $x^T x = 1$, (4.67) gives

$$0 < \underline{\lambda} x^T x \leqslant x^T B^*(u) x \leqslant \overline{\lambda} x^T x\,, \quad u \in [0, t]\,. \tag{4.68}$$

Finally, since (4.68) is homogeneous in x, (4.66) holds for all $x \in \mathbb{R}^M$. □

Let us now return to the observation process Z in (4.62) and write

$$R(u) = H(u) X(u) \tag{4.69}$$

$$Y(s) = \int_0^s R(u)\,du\,, \quad s \in [0, t]\,. \tag{4.70}$$

Then the components of $\boldsymbol{R}$ are m.s. continuous and those of $\boldsymbol{Y}$ are m.s. continuously differentiable, whereas

$$\frac{d}{ds}\boldsymbol{Y}(s) = \boldsymbol{R}(s)\,, \quad s \in [0, t] \tag{4.71}$$
$$\boldsymbol{Y}(0) = \boldsymbol{0}\,.$$

Now, (4.62) has the form

$$\boldsymbol{Z}(s) = \boldsymbol{Y}(s) + \boldsymbol{W}^*(s)\,, \quad s \in [0, t]\,. \tag{4.72}$$

Clearly, $\boldsymbol{Z}$ is m.s. continuous on $[0, t]$ and $\boldsymbol{Z}(0) = \boldsymbol{0}$.

Let $L(\boldsymbol{Z}, t)$ be the linear hull of the set

$$\{Z_1(s_1), \ldots, Z_1(s_k), \ldots, Z_M(s_1), \ldots, Z_M(s_k); \quad s_1, \ldots, s_k \text{ in } [0, t]\}\,.$$

If $V \in L(\boldsymbol{Z}, t)$, the result in C above applies. Hence, there are elements $f_1, \ldots, f_M$ in $I([0, t])$ such that

$$\begin{aligned} V &= \int_0^t f_1(s)\,dZ_1(s) + \ldots + \int_0^t f_M(s)\,dZ_M(s) \\ &= \int_0^t \boldsymbol{f}(s)^T d\boldsymbol{Z}(s)\,, \quad \text{where} \end{aligned} \tag{4.73}$$

$$\boldsymbol{f}(s) = \begin{bmatrix} f_1(s) \\ \vdots \\ f_M(s) \end{bmatrix}, \quad s \in [0, t]\,. \tag{4.74}$$

Let $H(\boldsymbol{Z}, t)$ be the closure of $L(\boldsymbol{Z}, t)$ in $L_2(\Omega)$. If $X \in H(\boldsymbol{Z}, t)$, there is a Cauchy sequence $\{X_n\}_{n \in N}$ in $L(\boldsymbol{Z}, t)$ converging to X in m.s. According to (4.73), for each X_n there is an M vector $\boldsymbol{f}_n$ with components f_{jn}, $j = 1, \ldots, M$, in $I([0, t])$ such that

$$X_n = \int_0^t \boldsymbol{f}_n^T(s)\,d\boldsymbol{Z}(s) \tag{4.75}$$

and thus (also following from Theorem 4.10)

$$\begin{aligned} \|X_m - X_n\|_\Omega^2 &= \mathrm{E}\left\{\int_0^t [\boldsymbol{f}_m(s) - \boldsymbol{f}_n(s)]^T d\boldsymbol{Z}(s)\right\}^2 \\ &= \mathrm{E}\left\{\int_0^t [\boldsymbol{f}_m(s) - \boldsymbol{f}_n(s)]^T \boldsymbol{R}(s)\,ds + \int_0^t [\boldsymbol{f}_m(s) - \boldsymbol{f}_n(s)]^T d\boldsymbol{W}^*(s)\right\}^2 \\ &= A_{m,n} + B_{m,n} \to 0 \end{aligned} \tag{4.76}$$

as $m, n \to \infty$, where

$$A_{m,n} = \mathrm{E}\left\{\int_0^t [f_m(s)-f_n(s)]^T \boldsymbol{R}(s)\,ds\right\}^2 \tag{4.77}$$

$$B_{m,n} = \mathrm{E}\left\{\int_0^t [f_m(s)-f_n(s)]^T d\boldsymbol{W}^*(s)\right\}^2 . \tag{4.78}$$

Since $A_{m,n}$ and $B_{m,n}$ are nonnegative, we have

$$A_{m,n} \to 0 \quad \text{and} \tag{4.79}$$

$$B_{m,n} \to 0 \tag{4.80}$$

as $m, n \to \infty$. We also see that [Theorem 2.22 (a)]

$$A_{m,n} = \int_0^t du \int_0^t dv\,[f_m(u)-f_n(u)]^T E\{\boldsymbol{R}(u)\boldsymbol{R}^T(v)\}[f_m(v)-f_n(v)] \tag{4.81}$$

and that (with the aid of Theorem 2.35)

$$B_{m,n} = \int_0^t [f_m(s)-f_n(s)]^T B^*(s)[f_m(s)-f_n(s)]\,ds . \tag{4.82}$$

According to the lemma given above, there are numbers $\underline{\lambda}$ and $\overline{\lambda}$, $0 < \underline{\lambda} \leqslant \overline{\lambda}$, such that

$$\begin{aligned} &\underline{\lambda}\,[f_m(s)-f_n(s)]^T[f_m(s)-f_n(s)] \\ &\quad \leqslant [f_m(s)-f_n(s)]^T B^*(s)[f_m(s)-f_n(s)] \\ &\quad \leqslant \overline{\lambda}\,[f_m(s)-f_n(s)]^T[f_m(s)-f_n(s)] . \end{aligned} \tag{4.83}$$

Upon integrating (4.83) from 0 to t, we have

$$\begin{aligned} 0 &\leqslant \underline{\lambda}\left\{\sum_{j=1}^M \|f_{jm}-f_{jn}\|^2_{[0,t]}\right\} \\ &\leqslant B_{m,n} \\ &\leqslant \overline{\lambda}\left\{\sum_{j=1}^M \|f_{jm}-f_{jn}\|^2_{[0,t]}\right\} . \end{aligned} \tag{4.84}$$

Thus, $B_{m,n} \to 0$ if and only if

$$\|f_{jm}-f_{jn}\|_{[0,t]} \to 0 \tag{4.85}$$

as $m, n \to \infty, j = 1, \ldots, M$. Hence, for each j, $\{f_{jn}\}_{n\in\mathbb{N}}$ is a Cauchy sequence in $I([0, t])$ with limit f_j, say, in $L_2([0, t])$, i.e.,

$$f_{jn} \to f_j \tag{4.86}$$

in $L_2([0, t])$ as $n \to \infty, j = 1, \ldots, M$.

Conversely, if (4.85) holds, then $B_{m,n} \to 0$ and also $A_{m,n} \to 0$ on account of Theorem 4.6 and definitions (4.27, 28).

Applying the results of (D), we have

$$X = \lim_{n\to\infty} X_n = \lim_{n\to\infty} \int_0^t \boldsymbol{f}_n^T(s)\, d\boldsymbol{Z}(s) = \int_0^t \boldsymbol{f}^T(s)\, d\boldsymbol{Z}(s)\,. \tag{4.87}$$

Given $X \in H(\boldsymbol{Z}, t)$, the M vector function $\boldsymbol{f}$ satisfying (4.87) is unique. To show this, if $\boldsymbol{h}$ with components h_j in $L_2([0, t])$ also satisfies (4.87), then

$$\begin{aligned}
0 &= \int_0^t [\boldsymbol{f}(s) - \boldsymbol{h}(s)]^T d\boldsymbol{Z}(s) \\
\Rightarrow 0 &= \left\| \int_0^t [\boldsymbol{f}(s) - \boldsymbol{h}(s)]^T d\boldsymbol{Z}(s) \right\|_\Omega^2 \\
\Rightarrow 0 &= \mathrm{E}\left\{ \int_0^t [\boldsymbol{f}(s) - \boldsymbol{h}(s)]^T \boldsymbol{R}(s)\, ds \right\}^2 + \mathrm{E}\left\{ \int_0^t [\boldsymbol{f}(s) - \boldsymbol{h}(s)]^T d\boldsymbol{W}^*(s) \right\}^2 \\
\Rightarrow 0 &= \int_0^t [\boldsymbol{f}(s) - \boldsymbol{h}(s)]^T B^*(s) [\boldsymbol{f}(s) - \boldsymbol{h}(s)]\, ds \\
&\geqslant \underline{\lambda} \int_0^t [\boldsymbol{f}(s) - \boldsymbol{h}(s)]^T [\boldsymbol{f}(s) - \boldsymbol{h}(s)]\, ds \\
\Rightarrow 0 &= \underline{\lambda} \sum_{j=1}^M \|f_j - h_j\|_{[0,t]}^2\,.
\end{aligned}$$

Since $\underline{\lambda} > 0, f_j = h_j$ a.e., $j = 1, \ldots, M$. In proving the above, Theorems 4.6, 10 and definitions (4.27, 28) were used.

Conversely, if the M vector function $\boldsymbol{f}$ is such that its components f_j, $j = 1, \ldots, M$, are elements of $L_2([0, t])$, it is easily seen that

$$\int_0^t \boldsymbol{f}^T(s)\, d\boldsymbol{Z}(s)$$

is an element of $H(\boldsymbol{Z}, t)$.

The central result of the above discussion is stated below as a theorem.

Theorem 4.11. $X \in H(\boldsymbol{Z}, t)$ if and only if there is an M vector function $\boldsymbol{f}$ with components f_j in $L_2([0, t])$ such that

$$X = \int_0^t \boldsymbol{f}^T(s)\, d\boldsymbol{Z}(s)\,. \tag{4.88}$$

Given X, $\boldsymbol{f}$ is unique a.e. [4.6].

We emphasize that the success of this theorem depends strongly on the special Wiener-Lévy process $\boldsymbol{W}^*$ in the observations.

4.5.1 Supplementary Exercises

Exercise 4.5. Show that a finite-dimensional subspace of $L_2(\Omega)$ is closed.

Exercise 4.6. Prove the following theorem.
If $W : [0, T] \to L_2(\Omega)$ is the standard Wiener-Lévy process and if

$$\mathscr{F} : L_2([0, t]) \to L_2(\Omega) , \quad t \in (0, t]$$

is defined by

$$\forall f \in L_2([0, t]), \qquad \mathscr{F}f = \int_0^t f(s)\, dW(s) ,$$

then $\mathscr{F}$ establishes an isometry between $L_2([0, t])$ and $H(W, t)$. This theorem goes back to *Karhunen* [4.9].

Exercise 4.7. Prove the following theorem.
The mapping $\mathscr{F} : L_2([0, t]) \to H(Z, t)$ defined by

$$\forall f \in L_2([0, t]) , \qquad \mathscr{F}f = \int_0^t f(s)\, dZ(s)$$

establishes an isomorphism between the two spaces, and both $\mathscr{F}$ as well as $\mathscr{F}^{-1}$ are bounded.

4.6 The Wiener-Hopf Equation

We are now in a position to establish a set of basic equations in the construction of the Kalman-Bucy filter. Let $\boldsymbol{X}(t)$, $t \in [0, T]$, be the state vector of the dynamic system (3.51). As seen from the discussion in Sect. 4.1, and given the observations $\mathbf{Z}(t)$ at all $s \in [0, t]$, the Kalman-Bucy estimator of $\boldsymbol{X}(t)$ is the N vector $\hat{\boldsymbol{X}}(t)$ whose components $\hat{X}_i(t)$ are the orthogonal projections of the components $X_i(t)$ of $\boldsymbol{X}(t)$ onto the subspace $H(\mathbf{Z}, t)$ of $L_2(\Omega)$ (Fig. 4.3).

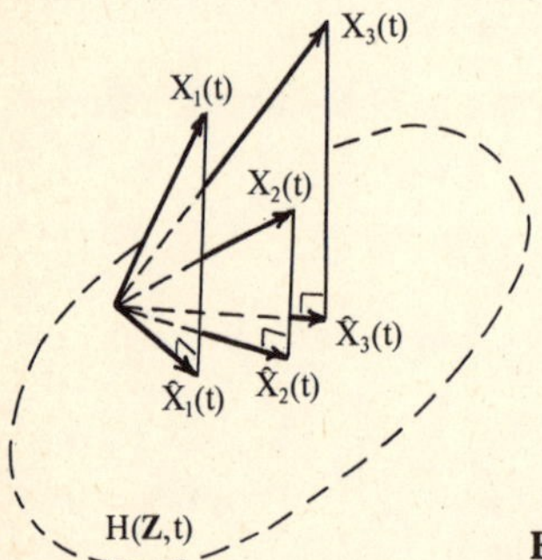

Fig. 4.3. Orthogonal projections of $X_i(t)$ onto $H(\mathbf{Z}, t)$

Equivalently, at each $i = 1,\ldots,N$, $\hat{X}_i(t)$ is required to satisfy

$$\hat{X}_i(t) \in H(\mathbf{Z}, t) \quad \text{and} \tag{4.89}$$

$$X_i(t) - \hat{X}_i(t) \in H^{\perp}(\mathbf{Z}, t) , \tag{4.90}$$

where $H^{\perp}(\mathbf{Z}, t)$ is the orthogonal complement of $H(\mathbf{Z}, t)$ in $L_2(\Omega)$.

For (4.89) to be valid, it is, according to Theorem 4.11, necessary and sufficient for $\hat{X}_i(t)$ to be represented by

$$\hat{X}_i(t) = \sum_{j=1}^{M} \int_0^t k_{ij}(t, u)\, dZ_j(u) , \quad i = 1,\ldots,N , \tag{4.91}$$

where $k_{ij}(t, u)$ belongs to $L_2([0, t])$ as a function of the variable u. In the above, $k_{ij}(t, u)$ is written as an explicit function of t as well, since t will not be a constant in what follows (cf. Sect. 4.4).

Let $K(t, u)$ be the $N \times M$ matrix with $k_{ij}(t, u)$ as its elements. Then we may write

$$\left.\begin{aligned} \hat{\mathbf{X}}(t) &= \int_0^t K(t, u)\, d\mathbf{Z}(u) \\ K(t, u) &= (k_{ij}(t,u))_{\substack{i=1,\ldots,N \\ j=1,\ldots,M}} , \quad k_{ij}(t,.) \in L_2([0, t]) \end{aligned}\right\} \tag{4.92}$$

and (4.92) is equivalent to (4.89).

For (4.90) to be valid, it is necessary and sufficient that, at each $s \in [0, t]$ and for each $j = 1,\ldots,M$, $X_i(t) - \hat{X}_i(t)$ is orthogonal to $Z_j(s)$ or

$$\forall j = 1,\ldots,M , \quad \forall s \in [0, t] , \quad \mathrm{E}\{[X_i(t) - \hat{X}_i(t)]\, Z_j(s)\} = 0 , \tag{4.93}$$

$i = 1,\ldots,$N. For then $X_i(t) - \hat{X}_i(t)$ is orthogonal to any element $U \in H(\mathbf{Z}, t)$ because for U there is a sequence $\{U_n\}_{n \in \mathbb{N}}$ converging to U in m.s. and whose elements U_n are linear combinations of $Z_j(s_k)$, $j = 1,\ldots,M$ and $s_k \in [0, t]$. Then,

$$\mathrm{E}\{[X_i(t) - \hat{X}_i(t)]\, U_n(s)\} = 0$$

and, due to continuity of the inner product in $L_2(\Omega)$,

$$\mathrm{E}\{[X_i(t) - \hat{X}_i(t)]\, U\} = 0 .$$

Hence, (4.90) is equivalent to

$$\mathrm{E}\{X_i(t)\, Z_j(s)\} = \mathrm{E}\{\hat{X}_i(t)\, Z_j(s)\} \tag{4.94}$$

at all $s \in [0, t]$, $j = 1,\ldots,M$, $i = 1,\ldots,N$, or to

$$\mathrm{E}\{\mathbf{X}(t)\, \mathbf{Z}^T(s)\} = \mathrm{E}\{\hat{\mathbf{X}}(t)\, \mathbf{Z}^T(s)\} \quad \text{at all} \quad s \in [0, t] . \tag{4.95}$$

Now, it is seen that $\hat{X}(t)$ is characterized by (4.92, 95). Equation (4.95) is the so-called *Wiener-Hopf* equation of the Kalman-Bucy filter.

In terms of its elements, (4.95) takes the form

$$\begin{bmatrix} \mathrm{E}\{X_1(t)\,Z_1(s)\} \dots \mathrm{E}\{X_1(t)\,Z_M(s)\} \\ \vdots \qquad\qquad \vdots \\ \mathrm{E}\{X_i(t)\,Z_1(s)\} \dots \mathrm{E}\{X_i(t)\,Z_M(s)\} \\ \vdots \qquad\qquad \vdots \\ \mathrm{E}\{X_N(t)\,Z_1(s)\} \dots \mathrm{E}\{X_N(t)\,Z_M(s)\} \end{bmatrix} = \begin{bmatrix} \mathrm{E}\{\hat{X}_1(t)\,Z_1(s)\} \dots \mathrm{E}\{\hat{X}_1(s)\,Z_M(s)\} \\ \vdots \qquad\qquad \vdots \\ \mathrm{E}\{\hat{X}_i(t)\,Z_1(s)\} \dots \mathrm{E}\{\hat{X}_i(t)\,Z_M(s)\} \\ \vdots \qquad\qquad \vdots \\ \mathrm{E}\{\hat{X}_N(t)\,Z_1(S)\} \dots \mathrm{E}\{\hat{X}_N(t)\,Z_M(s)\} \end{bmatrix}. \tag{4.96}$$

It is seen that $\hat{X}_i(t)$ is characterized by the M equations in the ith row of the above matrix equation and by (4.92). Hence, a decomposition of (4.92, 95) is possible, resulting in N subsystems of M equations each, characterizing $\hat{X}_i(t)$, $i = 1, \dots, N$.

Let us now examine what the Wiener-Hopf equation entails. Consider first the left-hand side of (4.95). Using [see (E) in Sect. 4.5]

$$\left.\begin{aligned} \boldsymbol{Z}(s) &= \boldsymbol{Y}(s) + \boldsymbol{W}^*(s) \\ \boldsymbol{Y}(s) &= \int_0^s \boldsymbol{R}(u)\,du \\ \boldsymbol{R}(u) &= H(u)\boldsymbol{X}(u) \end{aligned}\right\} \tag{4.97}$$

gives

$$\begin{aligned} \mathrm{E}\{\boldsymbol{X}(t)\boldsymbol{Z}^T(s)\} &= \mathrm{E}\{\boldsymbol{X}(t)[\boldsymbol{Y}(s) + \boldsymbol{W}^*(s)]^T\} \\ &= \mathrm{E}\{\boldsymbol{X}(t)\,\boldsymbol{Y}^T(s)\} \\ &= \int_0^s \mathrm{E}\{\boldsymbol{X}(t)\boldsymbol{R}^T(u)\}\,du\,. \end{aligned} \tag{4.98}$$

The second equality of the above follows since

$$\mathrm{E}\{\boldsymbol{X}(t)\,\boldsymbol{W}^{*T}(s)\} = 0$$

according to Theorem 4.10 owing to the orthogonality of the components of $\boldsymbol{X}$ and $\boldsymbol{W}^*$. The third equality results from

$$\mathrm{E}\{\boldsymbol{X}(t)\,\boldsymbol{Y}^T(s)\} = \mathrm{E}\left\{\boldsymbol{X}(t)\int_0^s \boldsymbol{R}^T(u)\,du\right\} = \int_0^s \mathrm{E}\{\boldsymbol{X}(t)\boldsymbol{R}^T(u)\}\,du\,.$$

This is seen by writing

$$\int_0^s \boldsymbol{R}^T(u)\,du$$

as a limit of Riemann sums followed by interchanging the limit and the expectation operations, which is allowed due to the continuity of the inner product in $L_2(\Omega)$.

The right-hand side of (4.95) gives

$$\mathrm{E}\{\hat{\boldsymbol{X}}(t)\boldsymbol{Z}^T(s)\} = \int_0^t K(t,u)\frac{\partial}{\partial u}\mathrm{E}\{\boldsymbol{Z}(u)\boldsymbol{Z}^T(s)\}\,du\,. \tag{4.99}$$

This result can be derived following the steps shown below:

$$\begin{aligned}
\mathrm{E}\{\hat{\boldsymbol{X}}(t)\boldsymbol{Z}^T(s)\} &= \mathrm{E}\left\{\left[\lim_{n\to\infty}\int_0^t K_n(t,u)\,d\boldsymbol{Z}(u)\right]\boldsymbol{Z}^T(s)\right\} \\
&= \lim_{n\to\infty}\mathrm{E}\left\{\left[\int_0^t K_n(t,u)\,d\boldsymbol{Z}(u)\right]\boldsymbol{Z}^T(s)\right\} \\
&= \lim_{n\to\infty}\int_0^t K_n(t,u)\,d_u\mathrm{E}\{\boldsymbol{Z}(u)\boldsymbol{Z}^T(s)\} \\
&= \lim_{n\to\infty}\int_0^t K_n(t,u)\frac{\partial}{\partial u}\mathrm{E}\{\boldsymbol{Z}(u)\boldsymbol{Z}^T(s)\}\,du \\
&= \int_0^t K(t,u)\frac{\partial}{\partial u}\mathrm{E}\{\boldsymbol{Z}(u)\boldsymbol{Z}^T(s)\}\,du\,,
\end{aligned}$$

where the entries of $K_n(t,\cdot)$ are elements of $I([0,t])$ [see (E) in Sect. 4.5]. In the above, each equality can be verified as follows: the first follows from an application of Theorem 4.11 and definition (4.26); the second is due to the continuity of the inner product in $L_2(\Omega)$; the third is a consequence of Theorem 2.20; the fourth follows from a well-known theorem of real analysis, contained in Theorem 2.27 by putting $L_2(\Omega) = \mathbb{R}$; the last equality is due to the continuity of the inner product in $L_2([0,t])$.

It remains to show the differentiability of $\mathrm{E}\{Z(u)\boldsymbol{Z}^T(s)\}$. Since the components of $\hat{\boldsymbol{W}}(s)$ are orthogonal to those of $\boldsymbol{Y}(u)$, we have

$$\begin{aligned}
\mathrm{E}\{\boldsymbol{Z}(u)\boldsymbol{Z}^T(s)\} &= \mathrm{E}\{[\boldsymbol{Y}(u)+\boldsymbol{W}^*(u)][\boldsymbol{Y}(s)+\boldsymbol{W}^*(s)]^T\} \\
&= E\{\boldsymbol{Y}(u)\boldsymbol{Y}^T(s)\} + \int_0^m B^*(v)\,dv\,, \quad m=\min(u,s)\,.
\end{aligned} \tag{4.100}$$

Since

$$\boldsymbol{Y}(u) = \int_0^u \boldsymbol{R}(v)\,dv\,,$$

it becomes

$$\mathrm{E}\{\boldsymbol{Z}(u)\boldsymbol{Z}^T(s)\} = \begin{cases} \int_0^u\int_0^s \mathrm{E}\{\boldsymbol{R}(v)\boldsymbol{R}^T(w)\}\,dv\,dw + \int_0^u B^*(v)\,dv, & 0\leqslant u\leqslant s \\[2ex] \int_0^u\int_0^s \mathrm{E}\{\boldsymbol{R}(v)\boldsymbol{R}^T(w)\}\,dv\,dw + \int_0^s B^*(v)\,dv, & s<u\leqslant t \end{cases}$$

and hence

$$\frac{\partial}{\partial u} \mathrm{E}\{\boldsymbol{Z}(u)\boldsymbol{Z}^T(s)\} = \begin{cases} \int\limits_0^s \mathrm{E}\{\boldsymbol{R}(u)\boldsymbol{R}^T(w)\}\,dw + B^*(u)\,, & 0 \leqslant u \leqslant s \\ \int\limits_0^s \mathrm{E}\{\boldsymbol{R}(u)\boldsymbol{R}^T(w)\}\,dw\,, & s < u \leqslant t \end{cases} . \tag{4.101}$$

Upon substituting (4.98, 99, 101) into (4.95), the Wiener-Hopf equation now takes the form, on account of the theorem of Fubini,

$$\begin{aligned} \int\limits_0^s \mathrm{E}\{\boldsymbol{X}(t)\boldsymbol{R}^T(u)\}\,du &= \int\limits_0^s K(t,u)\left\{\int\limits_0^s \mathrm{E}\{\boldsymbol{R}(u)\boldsymbol{R}^T(w)\}\,dw + B^*(u)\right\}du \\ &\quad + \int\limits_s^t K(t,u)\left\{\int\limits_0^s \mathrm{E}\{\boldsymbol{R}(u)\boldsymbol{R}^T(w)\}\,dw\right\}du \\ &= \int\limits_0^t K(t,u)\left[\int\limits_0^s \mathrm{E}\{\boldsymbol{R}(u)\boldsymbol{R}^T(w)\}\,dw\right]du \\ &\quad + \int\limits_0^s K(t,u)\,B^*(u)\,du \\ &= \int\limits_0^s \left[\int\limits_0^t K(t,u)\,\mathrm{E}\{\boldsymbol{R}(u)\boldsymbol{R}^T(w)\}\,du\right]dw \\ &\quad + \int\limits_0^s K(t,u)\,B^*(u)\,du\,. \end{aligned}$$

It can be rewritten as

$$\int\limits_0^s \left[K(t,u)\,B^*(u) + \int\limits_0^t K(t,w)\,\mathrm{E}\{\boldsymbol{R}(w)\boldsymbol{R}^T(u)\}\,dw - \mathrm{E}\{\boldsymbol{X}(t)\boldsymbol{R}^T(u)\}\right]du = 0 \tag{4.102}$$

at all $s \in [0, t]$. Hence, according to a property of Lebesgue integrals, the integrand of (4.102) is equal to zero at almost all $u \in [0, t]$, i.e.,

$$K(t,u)\,B^*(u) + \int\limits_0^t K(t,w)\,\mathrm{E}\{\boldsymbol{R}(w)\boldsymbol{R}^T(u)\}\,dw - \mathrm{E}\{\boldsymbol{X}(t)\boldsymbol{R}^T(u)\} = 0 \tag{4.103}$$

at almost all $u \in [0, t]$. The second and third terms of (4.103) are continuous in u due to the m.s. continuity of the components of $\boldsymbol{R}(u)$ and to the continuity of the inner product in $L_2(\Omega)$. Therefore, $K(t, u)\,B^*(u)$ can be identified with a matrix, a.e. identical to it, and continuous on $[0, t]$. Since $B^*(u)$ is continuous on $[0, t]$ and invertible, its inverse $B^{*-1}(u)$ is also continuous. The theorems stated below are thus verified.

Theorem 4.12. The elements of the $N \times M$ matrix $K(t, u)$ introduced in (4.92) are continuous in u on $[0, t]$.

Theorem 4.13. Matrix $K(t, u)$ satisfies the integral equation

$$K(t, s)\,B^*(s) + \int_0^t K(t, u)\,\mathrm{E}\{\boldsymbol{R}(u)\boldsymbol{R}^T(s)\}\,du = \mathrm{E}\{\boldsymbol{X}(t)\boldsymbol{R}^T(s)\} \qquad (4.104)$$

at all $s \in [0, t]$ and $t \in [0, T]$.

In the construction of the Kalman-Bucy filter, the matrix functions $B^*(s)$, $\mathrm{E}\{\boldsymbol{R}(u)\boldsymbol{R}^T(s)\}$ and $\mathrm{E}\{\boldsymbol{X}(t)\boldsymbol{R}^T(s)\}$ are understood to be given at any $s \in [0, t]$, $t \in [0, T]$. Thus, (4.104) may be seen as a system of MN integral equations with MN unknown functions $k_{ij}(t,\cdot)$. Since $B^*(s)$ is symmetric, transposition of (4.104) gives

$$B^*(s)\,K^T(t, s) + \int_0^t \mathrm{E}\{\boldsymbol{R}(s)\boldsymbol{R}^T(u)\}\,K^T(t, u)\,du = \mathrm{E}\{\boldsymbol{R}(s)\boldsymbol{X}^T(t)\} \qquad (4.105)$$

at all $s \in [0, t]$.

Denoting the ith column of $K^T(t, s)$ by $\boldsymbol{K}_i(t, s)$, the ith column of (4.105) yields

$$B^*(s)\,\boldsymbol{K}_i(t, s) + \int_0^t \mathrm{E}\{\boldsymbol{R}(s)\boldsymbol{R}^T(u)\}\,\boldsymbol{K}_i(t, u)\,du = \mathrm{E}\{X_i(t)\boldsymbol{R}(s)\},\ s \in [0, t] \qquad (4.106)$$

$i = 1,\ldots,N$. It is seen that at each i and at a fixed $t \in [0, T]$, (4.106) is a system of M coupled linear integral equations of the Fredholm type. As shown in the next section, they can be solved uniquely, giving Theorem 4.14.

Theorem 4.14. The Kalman-Bucy estimator of $\boldsymbol{X}(t)$ is $\hat{\boldsymbol{X}}(t)$ if and only if it can be represented as

$$\hat{\boldsymbol{X}}(t) = \int_0^t K(t, u)\,d\boldsymbol{Z}(u)\,, \qquad (4.107)$$

where $K(t, u)$ is the solution to (4.104).

The construction of the Kalman-Bucy estimator is thus complete. From the computational point of view, however, it is seen from (4.106) that, at different values of t, different sets of N systems of type (4.106) need to be constructed and solved, requiring large computer memory and computational labor. This is particularly undesirable since construction of the Kalman-Bucy estimators is often carried out in real time in practical applications. More expedient solution procedure is thus necessary. In Sect. 4.7, it is shown that system (4.104) may be transformed so that $\hat{\boldsymbol{X}}(t + h)$ may be computed recursively from the knowledge of $\hat{\boldsymbol{X}}(t)$ in a computationally efficient manner.

4.6.1 The Integral Equation (4.106)

Let us now consider (4.106). We begin with $B^*(s)$ by recalling that

$B^*(s)$ is symmetric

$B^*(s) > 0$ under an added condition and hence invertible

$B^*(s)$ is continuous on $[0, T]$.

According to the lemma in Sect. 4.5, there is a positive number λ such that

$$\boldsymbol{x}^T B^*(s)\boldsymbol{x} \geqslant \lambda \boldsymbol{x}^T \boldsymbol{x}\,, \quad \lambda > 0\,, \quad s \in [0, T]\,. \tag{4.108}$$

Since $B^*(s)$ is symmetric, there are orthogonal matrices $\Gamma(s)$ and real-valued functions $\lambda_1(s), \ldots, \lambda_M(s)$ such that

$$B^*(s) = \Gamma^T(s)\begin{bmatrix} \lambda_1(s) & & 0 \\ & \ddots & \\ 0 & & \lambda_M(s) \end{bmatrix}\Gamma(s) \tag{4.109}$$

and

$$\lambda_j(s) \geqslant \lambda > 0\,, \quad j = 1, \ldots, M\,, \quad s \in [0, T]\,. \tag{4.110}$$

Since $B^*(s)$ is continuous, the functions $\lambda_j(s)$ are also continuous and $\Gamma(s)$ may be considered to be continuous. Let us define

$$B^{*1/2}(s) = \Gamma^T(s)\begin{bmatrix} \lambda_1(s)^{1/2} & & 0 \\ & \ddots & \\ 0 & & \lambda_M(s)^{1/2} \end{bmatrix}\Gamma(s)\,. \tag{4.111}$$

Then the elements of $B^{*1/2}(s)$ are real valued on $[0, T]$ and possess the following properties:

$B^{*1/2}(s)\,B^{*1/2}(s) = B^*(s)$
$B^{*1/2}(s)$ is symmetric
$B^{*1/2}(s)$ is continuous
$B^{*1/2}(s) > 0$
$B^{*1/2}(s)$ is invertible.

Writing the inverse of $B^{*1/2}(s)$ as $B^{*-1/2}(s)$, then

$$B^{*1/2}(s)\,B^{*-1/2}(s) = I_M\,,$$

where I_M is the $M \times M$ identity matrix, and

$$B^{*-1/2}(s) = \Gamma^T(s) \begin{bmatrix} \lambda_1(s)^{-1/2} & & 0 \\ & \ddots & \\ 0 & & \lambda_M(s)^{-1/2} \end{bmatrix} \Gamma(s) . \tag{4.112}$$

The matrix $B^{*-1/2}(s)$ enjoys the same properties as $B^{*1/2}(s)$ stated above.

Premultiplying (4.106) by $B^{*-1/2}(s)$ gives

$$\begin{aligned} &B^{*1/2}(s)\boldsymbol{K}_i(t, s) + \int_0^t B^{*-1/2}(s)\,\mathrm{E}\{\boldsymbol{R}(s)\boldsymbol{R}^T(u)\}\, B^{*-1/2}(u)\, B^{*1/2}(u)\boldsymbol{K}_i(t, u)\, du \\ &\quad = B^{*-1/2}(s)\,\mathrm{E}\{X_i(t)\boldsymbol{R}(s)\} . \end{aligned} \tag{4.113}$$

If we define

$$\left.\begin{aligned} C(s, u) &= B^{*-1/2}(s)\,\mathrm{E}\{\boldsymbol{R}(s)\boldsymbol{R}^T(u)\}\, B^{*-1/2}(u) \\ \boldsymbol{\xi}_i(t, s) &= B^{*1/2}(s)\boldsymbol{K}_i(t, s) \\ \boldsymbol{\eta}_i(t, s) &= B^{*-1/2}(s)\,\mathrm{E}\{X_i(t)\boldsymbol{R}(s)\} \end{aligned}\right\} \tag{4.114}$$

s, t, and u in $[0, T]$, Eq. (4.113) reduces to

$$\boldsymbol{\xi}_i(t, s) + \int_0^t C(s, u)\,\boldsymbol{\xi}_i(t, u)\, du = \boldsymbol{\eta}_i(t, s) , \tag{4.115}$$

where, as seen from (4.114) $C(s, u)$ and $\boldsymbol{\eta}_i(t, s)$ are known functions and $\boldsymbol{\xi}_i(t, s)$ is to be determined.

Solving (4.106) is equivalent to solving (4.115). At each i, $i = 1, \ldots, N$, (4.115) is a Fredholm system if t is fixed in $[0, T]$; the kernel $C(s, u)$ is independent of i, $i = 1, \ldots, N$, and continuous in (s, u) on $[0, t]^2$ (Sect. 4.6). To proceed, let (Fig. 4.4)

$$D = \{(t, s) \mid 0 \leqslant s \leqslant t \leqslant T\} . \tag{4.116}$$

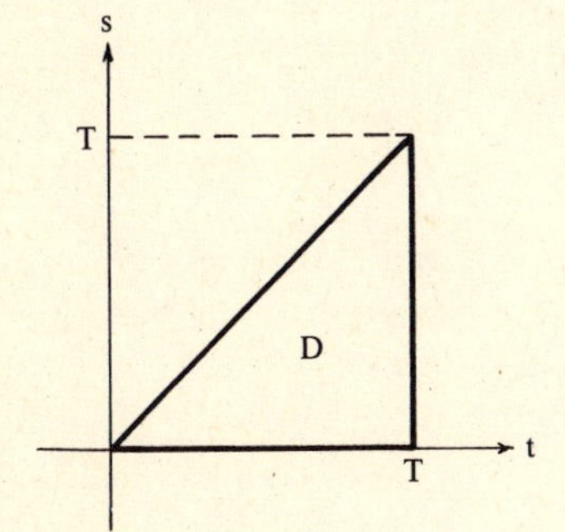

Fig. 4.4. The domain D

We shall show that $\boldsymbol{\eta}_i(t, s)$ is

a) continuous in (t, s) on D, and
b) partially differentiable with respect to t in D with partial derivative $\partial \boldsymbol{\eta}_i(t, s)/\partial t$ continuous in (t, s) on D.

Assertion (a) is true by virtue of continuity of $B^{*-1/2}(s)$, the m.s. continuity of $X_i(t)$ and the components of $\boldsymbol{R}(s)$, and the continuity of the inner product in $L_2(\Omega)$ (Sect. 4.6).

To prove assertion (b), it suffices to show that $\mathrm{E}\{\boldsymbol{R}(s)\boldsymbol{X}^T(t)\}$ is continuously differentiable with respect to t, where $\boldsymbol{X}(t)$ is the state vector of the dynamic system (3.51). Since

$$\frac{d}{dt}\boldsymbol{X}(t) = A(t)\boldsymbol{X}(t) ,$$

then (Chap. 3 and Theorem 2.14)

$$\frac{\partial}{\partial t}\mathrm{E}\{\boldsymbol{R}(s)\boldsymbol{X}^T(t)\} = \mathrm{E}\{\boldsymbol{R}(s)\boldsymbol{X}^T(t)\}A^T(t) .$$

Assertion (b) follows because of continuity of the inner product in $L_2(\Omega)$.

Since assertions (a) and (b) are true at each i, $i = 1, \ldots, N$, we shall drop this subscript in (4.115) and write

$$\boldsymbol{\xi}(t, s) + \int_0^t C(s, u)\,\boldsymbol{\xi}(t, u)\,du = \boldsymbol{\eta}(t, s) , \tag{4.117}$$

where $\boldsymbol{\eta}(t, s)$ has the properties (a) and (b).

We shall now show that the solution $\boldsymbol{\xi}(t, s)$ in (4.117) also has the properties (a) and (b). To do so, it is fruitful to work first in the space

$$L_2^M([0, t]) = \left\{\boldsymbol{a}(\cdot) = \begin{bmatrix} a_1(\cdot) \\ \vdots \\ a_M(\cdot) \end{bmatrix} \middle|\, a_j(\cdot) \in L_2([0, t]), j = 1, \ldots, M\right\} ,$$

t fixed in $[0, T]$. The usual rules of addition and multiplication by a scalar apply and an inner product is defined by

$$(\boldsymbol{a}(\cdot), \boldsymbol{b}(\cdot))_t = \sum_{j=1}^M \int_0^t a_j(u)\,b_j(u)\,du = \int_0^t \boldsymbol{a}^T(u)\,\boldsymbol{b}(u)\,du , \tag{4.118}$$

$$\boldsymbol{a}(\cdot) \text{ and } \boldsymbol{b}(\cdot) \text{ in } L_2^M([0, t]) .$$

Since $L_2([0, t])$ is a Hilbert space, $L_2^M([0, t])$ is also a Hilbert space provided that the elements whose respective components are a.e. identical in pairs on $[0, t]$ are identified. Under this restriction,

$$\|\boldsymbol{a}(\cdot)\|_t = [(\boldsymbol{a}(\cdot), \boldsymbol{a}(\cdot))_t]^{1/2} = \left[\sum_{j=1}^M \int_0^t a_j^2(u)\,du\right]^{1/2} \tag{4.119}$$

is a norm in $L_2^M([0, t])$. The Cauchy inequality in this context reads

$$|(\boldsymbol{a}(\cdot), \boldsymbol{b}(\cdot))_t| \leqslant \|\boldsymbol{a}(\cdot)\|_t \|\boldsymbol{b}(\cdot)\|_t . \tag{4.120}$$

Let $c_{jk}(s, u)$, j, $k = 1, \ldots, M$, be the elements of $C(s, u)$ and let

$$\boldsymbol{c}_j(s, u) = (c_{j1}(s, u), \ldots, c_{jM}(s, u))$$

be the jth row of $C(s, u)$. Since the elements of $C(s, u)$ are continuous in s and u on $[0, t]$, $\boldsymbol{c}_j(s,\cdot)^T$ may and will occasionally be looked upon as an element of $L_2^M([0, t])$.

For any $\boldsymbol{a}(\cdot) \in L_2^M([0, t])$, we consider

$$\boldsymbol{b}(t, s) = \int_0^t C(s, u)\boldsymbol{a}(u)\,du , \quad s \in [0, t] . \tag{4.121}$$

It is easily seen that $\boldsymbol{b}(t, s)$ is continuous in s on $[0, t]$, hence $\boldsymbol{b}(t,\cdot) \in L_2^M([0, t])$. Define now the mapping

$$\mathscr{C}_t : L_2^M([0, t]) \rightarrow L_2^M([0, t]) \quad \text{by}$$

$$\forall \boldsymbol{a}(\cdot) \in L_2^M([0, t]) : (\mathscr{C}_t \boldsymbol{a}(\cdot))(s) = \int_0^t C(s, u)\boldsymbol{a}(u)\,du, \; s \in [0, t] . \tag{4.122}$$

It possesses the following properties on the basis of some theorems on operators in Hilbert spaces [4.5].

Theorem 4.15. The operator $\mathscr{C}_t$ defined in (4.122) enjoys among others the following properties:

a) it is linear,
b) it is bounded (continuous) on $L_2^M([0, t])$ with norm $|\mathscr{C}_t|$ uniformly bounded in $t \in [0, t]$,
c) it is positive semidefinite,
d) it is a symmetric operator.

Proof (a). If $\boldsymbol{a}(\cdot)$ and $\boldsymbol{b}(\cdot)$ are elements of $L_2^M([0, t])$ and λ and μ are real numbers, then

$$(\mathscr{C}_t[\lambda \boldsymbol{a}(\cdot) + \mu \boldsymbol{b}(\cdot)])(s) = \int_0^t C(s, u)[\lambda \boldsymbol{a}(u) + \mu \boldsymbol{b}(u)]\,du$$

$$= \lambda \int_0^t C(s, u)\boldsymbol{a}(u)\,du + \mu \int_0^t C(s, u)\boldsymbol{b}(u)\,du$$

$$= [\lambda \mathscr{C}_t \boldsymbol{a}(\cdot) + \mu \mathscr{C}_t \boldsymbol{b}(\cdot)](s) .$$

b) Let

$$C = \left[\sum_{j=1}^{M} \sum_{k=1}^{M} \int_0^T \int_0^T c_{jk}^2(s, u)\,ds\,du\right]^{1/2} ,$$

where $c_{jk}(s, u)$ are (continuous) elements of $C(s, u)$. Then

$$\begin{aligned}
\|\mathscr{C}_t \boldsymbol{a}(\cdot)\|_t^2 &= \sum_{j=1}^{M} \int_0^t \left\{ \int_0^t \sum_{k=1}^{M} c_{jk}(s, u)\, a_k(u)\, du \right\}^2 ds \\
&= \sum_{j=1}^{M} \int_0^t (\boldsymbol{c}_j^T(s,\cdot), \boldsymbol{a}(\cdot))_t^2\, ds \\
&\leqslant \sum_{j=1}^{M} \int_0^t \|\boldsymbol{c}_j^T(s,\cdot)\|_t^2\, \|\boldsymbol{a}(\cdot)\|_t^2\, ds \\
&= \|\boldsymbol{a}(\cdot)\|_t^2 \sum_{j=1}^{M} \int_0^t \|\boldsymbol{c}_j^T(s,\cdot)\|_t^2\, ds \\
&= \|\boldsymbol{a}(\cdot)\|_t^2 \sum_{j=1}^{M} \sum_{k=1}^{M} \int_0^t \int_0^t c_{jk}^2(s, u)\, du\, ds \\
&\leqslant \|\boldsymbol{a}(\cdot)\|_t^2 \sum_{j=1}^{M} \sum_{k=1}^{M} \int_0^T \int_0^T c_{jk}^2(s, u)\, ds\, du \\
&= \|\boldsymbol{a}(\cdot)\|_t^2\, C^2 .
\end{aligned}$$

Hence,

$$\|\mathscr{C}_t \boldsymbol{a}(\cdot)\|_t \leqslant C \|\boldsymbol{a}(\cdot)\|_t$$

for all $\boldsymbol{a}(\cdot) \in L_2^M([0, t])$ and for all $t \in [0, T]$. Thus, $\mathscr{C}_t$ is bounded with norm $\|\mathscr{C}_t\|$ satisfying $\|\mathscr{C}_t\| \leqslant C$ uniformly in t on $[0, T]$.

c) This property follows because, for each $\boldsymbol{a}(\cdot) \in L_2^M([0, t])$,

$$\begin{aligned}
(\boldsymbol{a}(\cdot),\ \mathscr{C}_t \boldsymbol{a}(\cdot))_t &= \int_0^t \boldsymbol{a}^T(s) \int_0^t C(s, u)\, \boldsymbol{a}(u)\, du\, ds \\
&= \int_0^t \int_0^t \boldsymbol{a}^T(s)\, B^{*-1/2}(s)\, \mathrm{E}\{\boldsymbol{R}(s)\boldsymbol{R}^T(u)\}\, B^{*-1/2}(u)\, \boldsymbol{a}(u)\, du\, ds \\
&= \mathrm{E}\left\{ \int_0^t \int_0^t [\boldsymbol{a}^T(s)\, B^{*-1/2}(s)\, \boldsymbol{R}(s)]\, [\boldsymbol{a}^T(u)\, B^{*-1/2}(u)\, \boldsymbol{R}(u)]^T du\, ds \right\} \\
&= \mathrm{E}\left\{ \int_0^t \boldsymbol{a}^T(s)\, B^{*-1/2}(s)\, \boldsymbol{R}(s)\, ds \right\}^2 \geqslant 0 .
\end{aligned}$$

d) For all $\boldsymbol{a}(\cdot)$ and $\boldsymbol{b}(\cdot)$ in $L_2^M([0, t])$, we have

$$\begin{aligned}
(\boldsymbol{a}(\cdot),\ \mathscr{C}_t \boldsymbol{b}(\cdot))_t &= \int_0^t \boldsymbol{a}^T(s) \int_0^t C(s, u)\, \boldsymbol{b}(u)\, du\, ds \\
&= \int_0^t \int_0^t \boldsymbol{a}^T(s)\, C(s, u)\, \boldsymbol{b}(u)\, du\, ds \\
&= \int_0^t \int_0^t [\boldsymbol{a}^T(s)\, C(s, u)\, \boldsymbol{b}(u)]^T du\, ds . \qquad (4.123)
\end{aligned}$$

Since

$$\begin{aligned} C^T(s, u) &= [B^{*-1/2}(s)\, \mathrm{E}\{\boldsymbol{R}(s)\boldsymbol{R}^T(u)\}\, B^{*-1/2}(u)]^T \\ &= B^{*-1/2}(u)\, \mathrm{E}\{\boldsymbol{R}(u)\boldsymbol{R}^T(s)\}\, B^{*-1/2}(s) \\ &= C(u, s)\ , \end{aligned}$$

(4.123) yields

$$\begin{aligned} (\boldsymbol{a}(\cdot), \mathscr{C}_t \boldsymbol{b}(\cdot))_t &= \int_0^t \int_0^t \boldsymbol{b}^T(u)\, C(u, s)\, \boldsymbol{a}(s)\, ds\, du \\ &= (\boldsymbol{b}(\cdot), \mathscr{C}_t \boldsymbol{a}(\cdot))_t\ . \end{aligned}$$

The operator $\mathscr{C}_t$ is thus symmetric on $L_2^M([0, t])$. □

According to (d), the spectrum σ_t of $\mathscr{C}_t$ is contained in $\mathbb{R}$ and, because of (c), in $[0, \infty)$ (It may be proven that σ_t is contained in $[0, C]$ for all $t \in [0, T]$.)

If $\mathscr{I}_t : L_2^M([0, t]) \to L_2^M([0, t])$ is the identical mapping, then, according to the definition of the spectrum of an operator, the mapping

$$(\mathscr{C}_t - \lambda \mathscr{I}_t) : L_2^M([0, t]) \to L_2^M([0, t])$$

is invertible with a continuous inverse if $\lambda \notin \sigma_t$, i.e.,

$$(\mathscr{C}_t - \lambda \mathscr{I}_t)^{-1} : L_2^M([0, t]) \to L_2^M([0, t])$$

is defined and bounded if $\lambda \notin \sigma_t$. In particular, since $-1 \notin \sigma_t$, $(\mathscr{C}_t + \mathscr{I}_t)^{-1}$ is defined on all of $L_2^M([0, t])$ and bounded.

In terms of the operators $\mathscr{C}_t$ and $\mathscr{I}_t$, (4.117) may now be written as

$$(\mathscr{C}_t + \mathscr{I}_t)\, \boldsymbol{\xi}(t, \cdot) = \boldsymbol{\eta}(t, \cdot) \tag{4.124}$$

and the results given above have shown that (4.124) has a unique solution

$$\boldsymbol{\xi}(t, \cdot) = (\mathscr{C}_t + \mathscr{I}_t)^{-1}\, \boldsymbol{\eta}(t, \cdot) \tag{4.125}$$

in $L_2^M([0, t])$.

Theorem 4.16. For all $t \in [0, T]$

$$\|(\mathscr{C}_t + \mathscr{I}_t)^{-1}\| \leqslant 1\ . \tag{4.126}$$

Proof. Since, according to Theorem 4.15, $\mathscr{C}_t$ is a positive semidefinite operator, we have (see 4.4 e.g.)

$$(\mathscr{C}_t \boldsymbol{a}(\cdot), \boldsymbol{a}(\cdot))_t \geqslant 0$$

for all $\boldsymbol{a}(\cdot) \in L_2^M([0, t])$. Hence,

$$\begin{aligned}\|\boldsymbol{a}(\cdot)\|_t^2 &= (\boldsymbol{a}(\cdot), \boldsymbol{a}(\cdot))_t \\ &\leqslant (\mathscr{C}_t\boldsymbol{a}(\cdot), \boldsymbol{a}(\cdot))_t + (\boldsymbol{a}(\cdot), \boldsymbol{a}(\cdot))_t \\ &= (\mathscr{C}_t\boldsymbol{a}(\cdot) + \boldsymbol{a}(\cdot), \boldsymbol{a}(\cdot))_t \\ &= ((\mathscr{C}_t + \mathscr{I}_t)\boldsymbol{a}(\cdot), \boldsymbol{a}(\cdot))_t\,.\end{aligned}$$

Cauchy's inequality then gives

$$\|\boldsymbol{a}(\cdot)\|_t^2 \leqslant \|(\mathscr{C}_t + \mathscr{I}_t)\boldsymbol{a}(\cdot)\|_t\|\boldsymbol{a}(\cdot)\|_t\,.$$

Thus,

$$\|\boldsymbol{a}(\cdot)\|_t \leqslant \|(\mathscr{C}_t + \mathscr{I}_t)\boldsymbol{a}(\cdot)\|_t \tag{4.127}$$

for all $\boldsymbol{a}(\cdot) \in L_2^M([0, t])$. Setting

$$(\mathscr{C}_t + \mathscr{I}_t)\boldsymbol{a}(\cdot) = \boldsymbol{b}(\cdot)\,, \tag{4.128}$$

since $\mathscr{C}_t + \mathscr{I}_t$ is invertible, it is surjective and hence $\boldsymbol{b}(\cdot)$ runs through all of $L_2^M([0, t])$ if $\boldsymbol{a}(\cdot)$ does. Equations (4.127, 128) therefore give

$$\|(\mathscr{C}_t + \mathscr{I}_t)^{-1}\boldsymbol{b}(\cdot)\|_t \leqslant \|\boldsymbol{b}(\cdot)\|_t$$

for all $\boldsymbol{b}(\cdot) \in L_2^M([0, t])$. The assertion is thus proved. □

Corollary. The unique solution (4.125) satisfies

$$\|\boldsymbol{\xi}(t,\cdot)\|_t \leqslant \|\boldsymbol{\eta}(t,\cdot)\|_t\,. \tag{4.129}$$

We have so far derived some properties of $\boldsymbol{\xi}(t,\cdot)$ as an element of $L_2^M([0, t])$. But we need properties of its components $\xi_j(t,\cdot)$ as ordinary real functions. It is seen from Theorem 4.12 that $\boldsymbol{\xi}(t, s)$ is continuous in s if $\boldsymbol{\eta}(t, s)$ is continuous in s. We therefore may consider $\boldsymbol{\xi}(t,\cdot)$ and $\boldsymbol{\eta}(t,\cdot)$ as elements of the Banach space $C^M([0, t])$, i.e., the vector space of elements

$$\boldsymbol{a}(\cdot) = \begin{bmatrix} a_1(\cdot) \\ \vdots \\ a_M(\cdot) \end{bmatrix}$$

whose components are continuous real functions on $[0, t]$ and with norm $|\boldsymbol{a}(\cdot)|_t$ defined by

$$|\boldsymbol{a}(\cdot)|_t = \max_{\substack{s \in [0,t] \\ j=1,\dots,M}} |a_j(s)|\,. \tag{4.130}$$

The notation $\boldsymbol{a}(\cdot) \to \boldsymbol{a}_o(\cdot)$ in $C^M([0, t])$ is used to mean that each component $a_j(s)$ of $\boldsymbol{a}(\cdot)$ converges to the corresponding component $a_{oj}(s)$ of $\boldsymbol{a}_o(\cdot)$, uniformly in s on $[0, t]$.

Clearly,

$$\boldsymbol{a}(\cdot) \in C^M([0, t]) \Rightarrow \boldsymbol{a}(\cdot) \in L_2^M([0, t]) .$$

It is seen that

$$C^M([0, t]) \subset L_2^M([0, t])$$

or that $C^M([0, t])$ is a linear subspace of $L_2^M([0, t])$ and that it is closed and complete with respect to its norm.

Let us consider the restriction $\overline{\mathscr{C}}_t$ on $C^M([0, t])$ of the operator $\mathscr{C}_t$ on $L_2^M([0, t])$. It is easily seen that

$$\overline{\mathscr{C}}_t : C^M([0, t]) \to C^M([0, t]) .$$

It is well known that $\overline{\mathscr{C}}_t$ is a compact and hence bounded operator, and that its spectrum consists of a countable set of eigenvalues only, possibly with the exceptional value zero [4.7]. Since the eigenvalues of $\overline{\mathscr{C}}_t$ are eigenvalues of $\mathscr{C}_t$ as well, the spectrum $\bar{\sigma}_t$ of $\overline{\mathscr{C}}_t$, like σ_t, is contained in $[0, \infty)$. If

$$\overline{\mathscr{I}}_t : C^M([0, t]) \to C^M(0, t])$$

is the identical mapping, then, according to the definition of the spectrum of an operator and by means of arguments analogous to those about $\mathscr{C}_t$ and $\mathscr{I}_t$, $\overline{\mathscr{C}}_t + \overline{\mathscr{I}}_t$ has a continuous inverse

$$(\overline{\mathscr{C}}_t + \overline{\mathscr{I}}_t)^{-1} : C^M([0, t]) \to C^M([0, t]) .$$

We have thus proved the following.

Theorem 4.17. If the $M \times M$ matrix $C(s, u)$ is continuous on $[0, t]^2$ and if $\boldsymbol{\eta}(t, s)$ is continuous on domain D defined in (4.116), then (4.117) admits a unique solution $\boldsymbol{\xi}(t, s)$ in $C^M([0, t])$.

Moreover, if $\boldsymbol{\eta}_o(t, s)$ is also continuous on D and if $\boldsymbol{\xi}_o(t, s)$ is the (unique) M vector satisfying

$$\boldsymbol{\xi}_o(t, s) + \int_0^t C(s, u)\, \boldsymbol{\xi}_o(t, u)\, du = \boldsymbol{\eta}_o(t, s) , \quad \text{then}$$

$$\boldsymbol{\eta}(t, s) \to \boldsymbol{\eta}_o(t, s) \text{ in } C^M([0, t]) \Leftrightarrow \boldsymbol{\xi}(t, s) \to \boldsymbol{\xi}_o(t, s) \text{ in } C^M([0, t]) . \tag{4.131}$$

In particular, since $\boldsymbol{\xi}(t, s) = \boldsymbol{0}$ when $\boldsymbol{\eta}(t, s) = \boldsymbol{0}$,

$$\boldsymbol{\eta}(t, s) \to \boldsymbol{0} \quad \text{in} \quad C^M([0, t]) \Leftrightarrow \boldsymbol{\xi}(t, s) \to \boldsymbol{0} \quad \text{in} \quad C^M([0, t]) . \tag{4.132}$$

Thus far, t has been kept fixed. We shall now also consider the behavior of $\boldsymbol{\xi}(t, s)$ as a function of t.

Theorem 4.18. If $\boldsymbol{\xi}(t, s)$ is the solution of (4.117) at any $t \in [0, T]$, then $|\boldsymbol{\xi}(t,\cdot)|_t$ is bounded on $[0, T]$ and so the components $\xi_j(t, s)$ of $\boldsymbol{\xi}(t, s)$ are bounded on the domain D, $j = 1, \ldots, M$.

Proof. With $\boldsymbol{c}_j(s, u)$ denoting the jth row of the $M \times M$ matrix $C(s, u)$, $j = 1, \ldots, M$, the jth component of

$$\int_0^t C(s, u)\, \boldsymbol{\xi}(t, u)\, du$$

may be seen as the inner product $(\boldsymbol{c}_j^T(s,\cdot), \boldsymbol{\xi}(t,\cdot))_t$ in $L_2^M([0, t])$. It follows from the Cauchy inequality in $L_2^M([0, t])$ and from Theorem 4.16 that

$$\begin{aligned}
\left| \int_0^t C(s, u)\, \boldsymbol{\xi}(t, u)\, du \right|_t &= \left| \begin{bmatrix} (\boldsymbol{c}_1^T(s,\cdot), \boldsymbol{\xi}(t,\cdot))_t \\ \vdots \\ (\boldsymbol{c}_M^T(s,\cdot), \boldsymbol{\xi}(t,\cdot))_t \end{bmatrix} \right|_t \\
&\leqslant \left| \begin{bmatrix} \|\boldsymbol{c}_1^T(s,\cdot)\|_t \|\boldsymbol{\xi}(t,\cdot)\|_t \\ \vdots \\ \|\boldsymbol{c}_M^T(s,\cdot)\|_t \|\boldsymbol{\xi}(t,\cdot)\|_t \end{bmatrix} \right|_t \\
&= \max_{\substack{s \in [0, T] \\ j = 1, \ldots, M}} \|\boldsymbol{c}_j(s,\cdot)\|_t \|\boldsymbol{\eta}(t,\cdot)\|_t \,.
\end{aligned}$$

Using (4.117), we thus have

$$\begin{aligned}
|\boldsymbol{\xi}(t,\cdot)|_t &\leqslant \left| \int_0^t C(\cdot, u)\, \boldsymbol{\xi}(t, u)\, du \right|_t + |\boldsymbol{\eta}(t,\cdot)|_t \\
&\leqslant \max_{\substack{s \in [0, T] \\ j = 1, \ldots, M}} \|\boldsymbol{c}_j(s,\cdot)\|_t \|\boldsymbol{\eta}(t,\cdot)\|_t + |\boldsymbol{\eta}(t,\cdot)|_t \,.
\end{aligned}$$

The uniform continuity on D of $C(s, u)$ and $\boldsymbol{\eta}(t, s)$ together with the inequality given above show the boundedness of $|\boldsymbol{\xi}(t,\cdot)|_t$ on $[0, T]$ and therefore the boundedness of its components $\xi_j(t, s)$ on D. □

Theorem 4.19. Let the $M \times M$ matrix $C(s, u)$ be continuous on $[0, T]^2$, and let the M vector $\boldsymbol{\eta}(t, s)$ be continuous on the domain D defined in (4.116) and partially differentiable with respect to t with derivative $\partial \boldsymbol{\eta}(t, s)/\partial t$ continuous on D. Then, if $\boldsymbol{\xi}(t, s)$ satisfies (4.117), it enjoys the following properties:

a) $\boldsymbol{\xi}(t, s)$ is continuous on D,

b) $\boldsymbol{\xi}(t, s)$ is partially differentiable with respect to t and $\partial\boldsymbol{\xi}(t, s)/\partial t$ satisfies

$$\frac{\partial}{\partial t}\boldsymbol{\xi}(t, s) + \int_0^t C(s, u)\frac{\partial}{\partial t}\boldsymbol{\xi}(t, u)\,du = \frac{\partial}{\partial t}\boldsymbol{\eta}(t, s) - C(s, t)\,\boldsymbol{\xi}(t, t)\,, \tag{4.133}$$

c) $\partial\boldsymbol{\xi}(t, s)/\partial t$ is continuous on D.

Proof. Suppose $0 \leqslant t \leqslant t' \leqslant T$, $s \in [0, t]$ and $s' \in [0, t']$ as shown in Fig. 4.5. Let ξ_j and η_j stand for the components of $\boldsymbol{\xi}$ and $\boldsymbol{\eta}$, respectively, and c_{jk} for the elements of C.

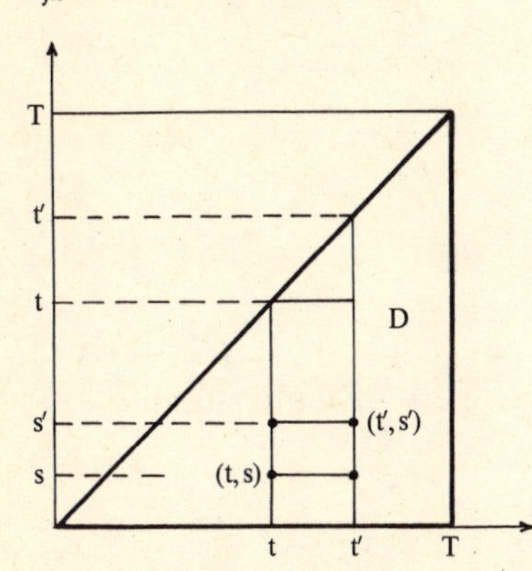

Fig. 4.5. $[0, T]^2$ and domain D

a) We shall show that, for any j, $j = 1, \ldots, M$,

$$|\xi_j(t', s') - \xi_j(t, s)|$$

is arbitrarily small if (t', s') is sufficiently close to (t, s). Consider

$$|\xi_j(t', s') - \xi_j(t, s)| \leqslant |\xi_j(t', s') - \xi_j(t', s)| + |\xi_j(t', s) - \xi_j(t, s)|\,. \tag{4.134}$$

According to (4.117), $\xi_j(t', s')$ and $\xi_j(t', s)$ satisfy

$$\xi_j(t', s') + \sum_{k=1}^{M}\int_0^{t'} c_{jk}(s', u)\,\xi_k(t', u)\,du = \eta_j(t', s') \quad \text{and} \tag{4.135}$$

$$\xi_j(t', s) + \sum_{k=1}^{M}\int_0^{t'} c_{jk}(s, u)\,\xi_k(t', u)\,du = \eta_j(t', s)\,. \tag{4.136}$$

In view of (4.135, 136), the first term on the right-hand side of (4.134) satisfies

$$|\xi_j(t', s') - \xi_j(t', s)| \leqslant |\eta_j(t', s') - \eta_j(t', s)| + \sum_{k=1}^{M}\int_0^{t'} |c_{jk}(s', u) - c_{jk}(s, u)|\,|\xi_k(t', u)|\,du\,.$$

Because of the uniform boundedness of $\xi_k(t', u)$ on D from Theorem 4.18 and the uniform continuity of $\eta_j(t', s)$ and $c_{jk}(s, u)$ in D, it is seen that $|\xi_j(t', s') - \xi_j(t, s)|$ is arbitrarily small, uniform in t, if $|s - s'|$ is sufficiently small.

Concerning the second term in (4.134), we first note that, again according to (4.117), $\boldsymbol{\xi}(t', s)$ and $\boldsymbol{\xi}(t, s)$ satisfy

$$\boldsymbol{\xi}(t', s) + \int_0^{t'} C(s, u)\,\boldsymbol{\xi}(t', u)\,du = \boldsymbol{\eta}(t', s) \qquad \text{and}$$

$$\boldsymbol{\xi}(t, s) + \int_0^{t} C(s, u)\,\boldsymbol{\xi}(t, u)\,du = \boldsymbol{\eta}(t, s)\,.$$

Hence, $\{\boldsymbol{\xi}(t', s) - \boldsymbol{\xi}(t, s)\}$ satisfies

$$[\boldsymbol{\xi}(t', s) - \boldsymbol{\xi}(t, s)] + \int_0^{t} C(s, u)[\boldsymbol{\xi}(t', u) - \boldsymbol{\xi}(t, u)]\,du$$

$$= \boldsymbol{\eta}(t', s) - \boldsymbol{\eta}(t, s) - \int_t^{t'} C(s, u)\,\boldsymbol{\xi}(t', u)\,du\,. \tag{4.137}$$

This is a relation in $[\boldsymbol{\xi}(t', s) - \boldsymbol{\xi}(t, s)]$ of the type given by (4.117). Because of uniform continuity of $\boldsymbol{\eta}(t, s)$ and the boundedness of the components of $C(s, u)\,\boldsymbol{\xi}(t', u)$ on D,

$$\left|\boldsymbol{\eta}(t',\cdot) - \boldsymbol{\eta}(t,\cdot) - \int_t^{t'} C(\cdot, u)\,\boldsymbol{\xi}(t', u)\,du\right|_t \to 0$$

as $t' \to t$. Hence, by virtue of (4.132), $|\boldsymbol{\xi}(t',\cdot) - \boldsymbol{\xi}(t,\cdot)|_t$ is arbitrarily small if $|t' - t|$ is sufficiently small, and the same is true of the components $|\xi_j(t', s) - \xi_j(t, s)|$, uniformly in s.

These results complete the proof of assertion (a).

b) From (4.137) we obtain

$$\frac{1}{t' - t}[\boldsymbol{\xi}(t', s) - \boldsymbol{\xi}(t, s)] + \int_0^{t} C(s, u)\frac{1}{t' - t}[\boldsymbol{\xi}(t', u) - \boldsymbol{\xi}(t, u)]\,du$$

$$= \frac{1}{t' - t}[\boldsymbol{\eta}(t', s) - \boldsymbol{\eta}(t, s)] - \frac{1}{t' - t}\int_t^{t'} C(s, u)\,\boldsymbol{\xi}(t', u)\,du\,. \tag{4.138}$$

We see that now

$$\frac{1}{t' - t}[\boldsymbol{\xi}(t', s) - \boldsymbol{\xi}(t, s)]$$

satisfies a relation of the same type as (4.117). Applying to the first term on the right-hand side of (4.138) the mean-value theorem of differential calculus and to the second the mean-value theorem of integral calculus, the right-hand side of (4.138) takes the form

$$\begin{bmatrix} \frac{\partial}{\partial t}\eta_1(\tau_1, s) \\ \vdots \\ \frac{\partial}{\partial t}\eta_M(\tau_M, s) \end{bmatrix} - \begin{bmatrix} c_{11}(s, \theta_1)\,\xi_1(t', \theta_1) & + \ldots + & c_{1M}(s, \theta_1)\,\xi_M(t', \theta_1) \\ & \vdots & \\ c_{M1}(s, \theta_M)\,\xi_1(t', \theta_M) & + \ldots + & c_{MM}(s, \theta_M)\,\xi_M(t', \theta_M) \end{bmatrix}$$

$$t < \tau_j < t', \quad t < \theta_j < t', \quad j = 1, \ldots, M . \tag{4.139}$$

We note that application of the mean-value theorems is permissible since, among others, $\boldsymbol{\xi}(t', u)$ is continuous on domain D.

Because of the uniform continuity on D of all functions in (4.139), the right-hand side of (4.138) converges to

$$\frac{\partial}{\partial t}\boldsymbol{\eta}(t, s) - C(s, t)\,\boldsymbol{\xi}(t, t)$$

as $t' \to t$, uniformly in s, i.e., in $C^M([0, t])$. On account of (4.131),

$$\frac{1}{t' - t}\,[\boldsymbol{\xi}(t', s) - \boldsymbol{\xi}(t, s)]$$

also converges in $C^M([0, t])$ as $t' \to t$ necessarily to

$$\frac{\partial}{\partial t}\boldsymbol{\xi}(t, s) .$$

Substituting these results into (4.138) and Theorem 4.17 establishes (4.133).

c) The proof is analogous to that of (a). □

As we rewrite the foregoing results in terms of the original variables [see (4.114)], we obtain the following theorem.

Theorem 4.20. The integral equation (4.104) has a unique solution $K(t, s)$ on the domain D defined in (4.116). Solution $K(t, s)$ is continuous on D and partially differentiable with respect to t; the partial derivative $\partial K(t, s)/\partial t$ is also continuous on D, satisfying

$$\frac{\partial}{\partial t}K(t, s)\,B^*(s) + \int_0^t \frac{\partial}{\partial t}K(t, u)\,\mathrm{E}\{\boldsymbol{R}(u)\boldsymbol{R}^T(s)\}\,du$$
$$= A(t)\,\mathrm{E}\{\boldsymbol{X}(t)\boldsymbol{R}^T(s)\} - K(t, t)\,\mathrm{E}\{\boldsymbol{R}(t)\boldsymbol{R}^T(s)\} . \tag{4.140}$$

The content of this section may be found in [4.7].

4.7 Kalman-Bucy Filter and the Riccati Equation

As pointed out in the preceding section, while a construction of the Kalman-Bucy estimator $\hat{X}(t)$ has been carried out, computationally expedient algorithms for determining $\hat{X}(t)$ are desired since it is required in real time in practical applications. This section addresses the development of an expedient procedure developed by Kalman and Bucy. Extensive references are available [4.8, 10]. Let us first summarize and recapitulate the relevant data and results obtained thus far.

Central to the discussion is a dynamic system with state vector $X(t)$ governed by

$$X(t) = C + \int_0^t A(s)X(s)\,ds + W(t), \quad t \in [0, T]. \tag{4.141}$$

In the above, $A(s)$ is a deterministic $N \times N$ matrix whose elements are continuous, real-valued functions of s on $[0, T]$; C is a zero-mean normally distributed stochastic N vector; $W(t)$ is a Wiener-Lévy N vector, stochastically independent of C and characterized by

$$\mathrm{E}W(t) = 0, \quad \mathrm{E}\{W(s)W^T(t)\} = \int_0^s B(u)\,du, \quad 0 \leq s \leq t \leq T.$$

The $N \times N$ matrix $B(u)$ has continuous, real-valued elements and is symmetric and positive semidefinite at each $t \in [0, T]$ (Sect. 2.6.1).

Let $\Phi(t)$ be the $N \times N$ fundamental matrix associated with $A(t)$, i.e.,

$$\frac{d}{dt}\Phi(t) = A(t)\Phi(t), \quad t \in [0, T], \quad \Phi(0) = I, \tag{4.142}$$

where I is the $N \times N$ identity matrix (Theorem 3.4). The solution to (4.141) may be represented as (Theorem 3.5)

$$X(t) = \Phi(t)\left[C + \int_0^t \Phi^{-1}(s)\,dW(s)\right], \quad t \in [0, T] \quad \text{with} \tag{4.143}$$

$$\mathrm{E}X(t) = 0$$

$$\mathrm{E}\{X(s)X^T(t)\} = \Phi(s)\left[\mathrm{E}\{CC^T\} + \int_0^s \Phi^{-1}(u)B(u)[\Phi^{-1}(u)]^T du\right]\Phi^T(t),$$

$$0 \leq s \leq t \leq T.$$

The variance $V(t) = \mathrm{E}\{X(t)X^T(t)\}$ satisfies (Theorem 3.5)

$$\frac{d}{dt}V(t) = A(t)V(t) + B(t) + V(t)A^T(t), t \in [0, T], V(0) = \mathrm{E}\{CC^T\}. \tag{4.144}$$

The state vector $X(t)$ is observed by

$$Z(s) = \int_0^s H(u)X(u)\,du + W^*(s)\,, \quad 0 \leq s \leq t \leq T\,, \tag{4.145}$$

where $H(u)$ is an $M \times N$ matrix with real-valued continuous elements defined on $[0, T]$; $W^*(s)$ is a Wiener-Lévy M vector, stochastically independent of C and $W(t)$ on $[0, T]$, and hence of $X(t)$, with

$$\mathrm{E}W^*(s) = 0\,, \quad \mathrm{E}\{W^*(s)W^{*T}(t)\} = \int_0^s B^*(u)\,du\,, \quad 0 \leq s \leq t \leq T\,.$$

The $M \times M$ matrix $B^*(u)$ is real valued, continuous, symmetric and necessarily positive semidefinite at each $u \in [0, T]$. Moreover, due to the extra condition

$$B^*(u) > 0$$

at each $u \in [0, T]$, $B^*(u)$ is also invertible with continuous inverse $B^{*-1}(u)$ at each $u \in [0, T]$ (Sect. 4.5).

The Kalman-Bucy estimator $\hat{X}(t)$ of $X(t)$ is an N vector whose components are orthogonal projections of the components of $X(t)$ onto the (closed) subspace $H(Z, t)$ of $L_2(\Omega)$ generated by the components of $Z(s)$ at all $s \in [0, t]$. We have introduced

$$\left.\begin{aligned} R(u) &= H(u)X(u) \\ Y(s) &= \int_0^s R(u)\,du \\ \text{and hence} & \\ Z(s) &= Y(s) + W^*(s)\,. \end{aligned}\right\} \tag{4.146}$$

The estimator $\hat{X}(t)$ may be represented by

$$\hat{X}(t) = \int_0^t K(t, u)\,dZ(u)\,, \tag{4.147}$$

where the elements of the $N \times M$ matrix $K(t, u)$ are naturally elements of $L_2([0, t])$ as functions of u.

In Sect. 4.6, it is shown that $K(t, s)$ is the (unique) solution to the matrix Fredholm system

$$K(t, s)B^*(s) + \int_0^t K(t, u)\,\mathrm{E}\{R(u)R^T(s)\}\,du = \mathrm{E}\{X(t)R^T(s)\}\,, \quad s \in [0, t]\,. \tag{4.148}$$

Defining domain D by

$$D = \{(t, s) \mid 0 \leq s \leq t \leq T\}\,, \tag{4.149}$$

it is shown in Sect. 4.6 that $K(t, s)$ is continuous in (t, s) on D and partially differentiable with respect to t and with derivative $\partial K(t, s)/\partial t$ also continuous in (t, s) on D. The derivative satisfies

$$\frac{\partial}{\partial t} K(t, s) B^*(s) + \int_0^t \frac{\partial}{\partial t} K(t, u) \, \mathrm{E}\{\boldsymbol{R}(u) \boldsymbol{R}^T(s)\} \, du$$

$$= A(t) \mathrm{E}\{\boldsymbol{X}(t) \boldsymbol{R}^T(s)\} - K(t, t) \mathrm{E}\{\boldsymbol{R}(t) \boldsymbol{R}^T(s)\} \, . \tag{4.150}$$

4.7.1 Recursion Formula and the Riccati Equation

Kalman and Bucy's recursion method of constructing $\hat{\boldsymbol{X}}(t)$ is now developed. We shall work with $K(s, s)$ instead of $K(t, s)$.

Since

$$K(t, s) = K(s, s) + \int_s^t \frac{\partial}{\partial u} K(u, s) \, du$$

we have from (4.147)

$$\hat{\boldsymbol{X}}(t) = \int_0^t K(t, s) \, d\boldsymbol{Z}(s)$$

$$= \int_0^t K(s, s) \, d\boldsymbol{Z}(s) + \int_0^t \left[\int_s^t \frac{\partial}{\partial u} K(u, s) \, du \right] \mathrm{d}\boldsymbol{Z}(s) \, . \tag{4.151}$$

Changing the order of integration gives

$$\hat{\boldsymbol{X}}(t) = \int_0^t K(s, s) \, d\boldsymbol{Z}(s) + \int_0^t \left[\int_0^u \frac{\partial}{\partial u} K(u, s) \, d\boldsymbol{Z}(s) \right] du \, . \tag{4.152}$$

The proof that interchanging the integration order is allowed is left as Exercise 4.8. From the first equation of (4.146) we have

$$\mathrm{E}\{\boldsymbol{R}(t) \boldsymbol{R}^T(s)\} = H(t) \mathrm{E}\{\boldsymbol{X}(t) \boldsymbol{R}^T(s)\} \quad \text{and}$$

$$K(t, t) \mathrm{E}\{\boldsymbol{R}(t) \boldsymbol{R}^T(s)\} = K(t, t) H(t) \mathrm{E}\{\boldsymbol{X}(t) \boldsymbol{R}^T(s)\} \, . \tag{4.153}$$

Substituting (4.153) into (4.150) yields

$$\frac{\partial}{\partial t} K(t, s) B^*(s) + \int_0^t \frac{\partial}{\partial t} K(t, u) \mathrm{E}\{\boldsymbol{R}(u) \boldsymbol{R}^T(s)\} \, du$$

$$= [A(t) - K(t, t) H(t)] \mathrm{E}\{\boldsymbol{X}(t) \boldsymbol{R}^T(s)\} \, . \tag{4.154}$$

Using (4.154), by premultiplying (4.148) by $[A(t) - K(t, t) H(t)]$ we get

$$\frac{\partial}{\partial t} K(t, s) = [A(t) - K(t, t) H(t)] K(t, s) \tag{4.155}$$

owing to uniqueness of the solution to (4.148).

Relations (4.152, 155) now lead to

$$\hat{X}(t) = \int_0^t K(s, s)\, dZ(s) + \int_0^t \left[\int_0^u [A(u) - K(u, u) H(u)] K(u, s)\, dZ(s) \right] du$$

and, with the aid of (4.147),

$$\hat{X}(t) = \int_0^t K(s, s)\, dZ(s) + \int_0^t [A(u) - K(u, u) H(u)] \hat{X}(u)\, du\,. \tag{4.156}$$

Equation (4.156) gives the desired recursion formula for constructing $\hat{X}(t)$ if $K(s, s)$, $s \in [0, T]$, is known since, with $t \leqslant t'$, $\hat{X}(t')$ can be recursively determined from

$$\hat{X}(t') = \hat{X}(t) + \int_t^{t'} K(s, s)\, dZ(s) + \int_t^{t'} [A(s) - K(s, s) H(s)] \hat{X}(s)\, ds\,. \tag{4.157}$$

$K(s, s)$ can be computed with the aid of a differential equation for the "error covariance matrix" $P(t)$ defined by

$$P(t) = \mathrm{E}\{\bar{X}(t) \bar{X}^T(t)\}\,, \quad \text{where} \tag{4.158}$$

$$\bar{X}(t) = X(t) - \hat{X}(t)\,. \tag{4.159}$$

Since the components of $\bar{X}(t)$ are orthogonal to the subspace $H(Z, t)$ of $L_2(\Omega)$ generated by the components of the observations $Z(s)$, $s \in [0, t]$, and since the components of $\hat{X}(t)$ belong to $H(Z, t)$, we have

$$\mathrm{E}\{\bar{X}(t) \hat{X}(t)\} = 0$$

and (4.158) becomes

$$P(t) = \mathrm{E}\{\bar{X}(t) X^T(t)\}\,. \tag{4.160}$$

Using the above and (4.146, 147), we get

$$\begin{aligned} P(t) H^T(t) &= \mathrm{E}\{\bar{X}(t) X^T(t)\} H^T(t) \\ &= \mathrm{E}\{[X(t) - \hat{X}(t)] R^T(t)\} \\ &= \mathrm{E}\{X(t) R^T(t)\} - \mathrm{E}\left\{\left[\int_0^t K(t, s)\, dZ(s)\right] R^T(t)\right\}. \end{aligned} \tag{4.161}$$

Calculus in m.s. and (4.146) show that

$$\left.\begin{aligned} Z(s) &= Y(s) + W^*(s) \\ \frac{d}{ds} Y(s) &= R(s)\,. \end{aligned}\right\} \tag{4.162}$$

Since the components of $\boldsymbol{W}^*(s)$ are orthogonal to those of $\boldsymbol{X}(t)$, we also have

$$\left.\begin{aligned} &\mathrm{E}\{\boldsymbol{W}^*(s)\boldsymbol{X}^T(t)\} = 0 \\ &\mathrm{E}\{\boldsymbol{W}^*(s)\boldsymbol{R}^T(t)\} = \mathrm{E}\{\boldsymbol{W}^*(s)\boldsymbol{X}^T(t)\}H^T(t) = 0\,. \end{aligned}\right\} \tag{4.163}$$

Again using m.s. calculus from Chap. 2, (4.162, 163) combine to give

$$\begin{aligned} &\mathrm{E}\left\{\left[\int_0^t K(t,s)\,d\boldsymbol{Z}(s)\right]\boldsymbol{R}^T(t)\right\} \\ &= \mathrm{E}\left\{\int_0^t K(t,s)\,d[\boldsymbol{Y}(s) + \boldsymbol{W}^*(s)]\boldsymbol{R}^T(t)\right\} \\ &= \mathrm{E}\left\{\left[\int_0^t K(t,s)\frac{d}{ds}\boldsymbol{Y}(s)\,ds\right]\boldsymbol{R}^T(t)\right\} + \int_0^t K(t,s)\,d\mathrm{E}\{\boldsymbol{W}^*(s)\boldsymbol{R}^T(t)\} \\ &= \int_0^t K(t,s)\,\mathrm{E}\{\boldsymbol{R}(s)\boldsymbol{R}^T(t)\}\,ds\,. \end{aligned}$$

The substitution of the above into (4.161) then gives

$$P(t)H^T(t) + \int_0^t K(t,s)\,\mathrm{E}\{\boldsymbol{R}(s)\boldsymbol{R}^T(t)\}\,ds = \mathrm{E}\{\boldsymbol{X}(t)\boldsymbol{R}^T(t)\}\,. \tag{4.164}$$

Comparing (4.164) with (4.148) we obtain, again because of uniqueness of the solution to (4.148),

$$P(t)H^T(t) = K(t,t)B^*(t) \quad \text{or}$$

$$K(t,t) = P(t)H^T(t)B^{*-1}(t) \tag{4.165}$$

since $B^*(t)$ is invertible.

It is seen that knowing $P(t)$ determines $K(t,t)$. We now shall show that the $N \times N$ matrix $P(t)$ is the solution of a matrix Riccati differential equation.

Equations (4.146) and m.s. calculus give

$$\begin{aligned} \int_0^t K(s,s)\,d\boldsymbol{Z}(s) &= \int_0^t K(s,s)\,d\left[\int_0^s \boldsymbol{R}(u)\,du + \boldsymbol{W}^*(s)\right] \\ &= \int_0^t K(s,s)H(s)\boldsymbol{X}(s)\,ds + \int_0^t K(s,s)\,d\boldsymbol{W}^*(s)\,. \end{aligned}$$

Returning to (4.156), it can thus be written as

$$\begin{aligned} \hat{\boldsymbol{X}}(t) = &\int_0^t K(s,s)H(s)\boldsymbol{X}(s)\,ds + \int_0^t K(s,s)\,d\boldsymbol{W}^*(s) \\ &+ \int_0^t [A(s) - K(s,s)H(s)]\hat{\boldsymbol{X}}(s)\,ds \\ = &\int_0^t K(s,s)H(s)\overline{\boldsymbol{X}}(s)\,ds + \int_0^t A(s)\hat{\boldsymbol{X}}(s)\,ds \\ &+ \int_0^t K(s,s)\,d\boldsymbol{W}^*(s)\,. \end{aligned}$$

Substracting the above from (4.141), we have

$$\begin{aligned}\overline{\boldsymbol{X}}(t) &= \boldsymbol{C} + \int_0^t [A(s) - K(s, s) H(s)] \overline{\boldsymbol{X}}(s)\, ds + \boldsymbol{W}(t) - \int_0^t K(s, s)\, d\boldsymbol{W}^*(s) \\ &= \boldsymbol{C} + \int_0^t \overline{A}(s) \overline{\boldsymbol{X}}(s)\, ds + \overline{\boldsymbol{W}}(t) \qquad (4.166)\end{aligned}$$

where we have used

$$\left.\begin{aligned}\overline{A}(t) &= A(t) - K(t, t) H(t) \\ \overline{\boldsymbol{W}}(t) &= \boldsymbol{W}(t) - \int_0^t K(s, s)\, d\boldsymbol{W}^*(s)\end{aligned}\right\} t \in [0, T]\,. \qquad (4.167)$$

In the above, $\overline{A}(t)$ is a continuous $N \times N$ matrix, $\boldsymbol{C}$ and $\overline{\boldsymbol{W}}(t)$ are stochastically independent and, as we shall show below, $\overline{\boldsymbol{W}}(t)$ is a Wiener-Lévy N vector.

To show this, we first see that $\overline{\boldsymbol{W}}(t)$ is normally distributed since its components are elements of the Gaussian space generated by the independent processes $\boldsymbol{W}(t)$ and $\boldsymbol{W}^*(s)$. The mean of $\overline{\boldsymbol{W}}(t)$ is

$$\mathrm{E}\overline{\boldsymbol{W}}(t) = \mathrm{E}\boldsymbol{W}(t) - \mathrm{E}\left\{\int_0^t K(s, s)\, \mathrm{d}\boldsymbol{W}^*(s)\right\} = \boldsymbol{0}$$

and, since $\mathrm{E}\{\boldsymbol{W}(s) \boldsymbol{W}^{*T}(t)\} = 0$ and due to Theorem 2.35,

$$\begin{aligned}\mathrm{E}\{\overline{\boldsymbol{W}}(s) \overline{\boldsymbol{W}}^T(t)\} &= \mathrm{E}\left\{\left[\boldsymbol{W}(s) - \int_0^s K(u, u)\, d\boldsymbol{W}^*(u)\right]\right. \\ &\qquad \left.\times \left[\boldsymbol{W}(t) - \int_0^t K(u, u)\, d\boldsymbol{W}^*(u)\right]^T\right\} \\ &= \int_0^s \overline{B}(u)\, du\,, \qquad (4.168)\end{aligned}$$

where

$$\overline{B}(u) = B(u) + K(u, u) B^*(u) K^T(u, u)\,, \quad u \in [0, T]\,. \qquad (4.169)$$

At each $u \in [0, T]$, the $N \times N$ matrix $\overline{B}(u)$ is symmetric, continuous and positive semidefinite. The last property is seen as follows. If $\boldsymbol{x}$ is any element of $\mathbb{R}^N$, then

$$\boldsymbol{y}(u) = K^T(u, u) \boldsymbol{x}$$

is also an element of $\mathbb{R}^N$. Hence, from (4.169),

$$\boldsymbol{x}^T \overline{B}(u) \boldsymbol{x} = \boldsymbol{x}^T B(u) \boldsymbol{x} + \boldsymbol{y}^T(u) B^*(u) \boldsymbol{y}(u) \geqslant 0$$

as both terms on the right-hand side are nonnegative at any $u \in [0, T]$.

We have thus shown that $\overline{W}(s)$, $s \in [0, T]$, has the properties of a general Wiener-Lévy N vector as defined in Sect. 2.6.1.

Now, (4.166) is of the same type as the differential equation (3.51) of Theorem 3.5. Hence, according to (3.54),

$$\left.\begin{aligned} \frac{d}{dt}P(t) &= \overline{A}(t)P(t) + \overline{B}(t) + P(t)\overline{A}^T(t)\,, \quad t \in [0, T] \\ P(0) &= \mathrm{E}\{\boldsymbol{C}\boldsymbol{C}^T\}\,. \end{aligned}\right\} \tag{4.170}$$

Reverting to the original functions using (4.165, 167, 169), (4,170) simplifies to

$$\left.\begin{aligned} \frac{d}{dt}P(t) &= B(t) + A(t)P(t) + P(t)A^T(t) \\ &\quad - P(t)H^T(t)B^{*-1}(t)H(t)P(t), \quad t \in [0, T] \\ P(0) &= \mathrm{E}\{\boldsymbol{C}\boldsymbol{C}^T\}\,. \end{aligned}\right\} \tag{4.171}$$

In the above, use has been made of the symmetry of the matrices $P(t)$ and $B^{*-1}(t)$.

Equation (4.171) is in the form of a matrix Riccati differential equation for $P(t)$ as we had set out to show. The determination of $P(t)$ from this equation gives $K(t, t)$ from (4.165), which in turn leads to the construction of $\hat{X}(t)$, the Kalman-Bucy estimator, from (4.156). To formalize this statement, some comments are in order.

Using (4.165), (4.156) may be written as

$$\begin{aligned} \hat{X}(t) &= \int_0^t [A(t) - P(s)H^T(s)B^{*-1}(s)H(s)]\hat{X}(s)\,ds \\ &\quad + \int_0^t P(s)H^T(s)B^{*-1}(s)\,d\boldsymbol{Z}(s)\,, \quad t \in [0, T]\,. \end{aligned} \tag{4.172}$$

The components of $\hat{X}(t)$ are unique orthogonal projections of the components of $X(t)$ onto the Hilbert space $H(\boldsymbol{Z}, t)$ and $X(t)$ is the unique solution of (4.141). Since

$$\overline{X}(t) = X(t) - \hat{X}(t)$$

satisfies (4.166) and its solution is unique, $\overline{X}(t)$ is the unique solution to (4.166).

Next, $P(t) = \mathrm{E}\{\overline{X}(t)\overline{X}^T(t)\}$ satisfies (4.171) globally, i.e., at each $t \in [0, T]$. Since solutions of equations of type (4.171) are unique, $P(t)$ is the unique global solution to (4.171), leading to Theorem 4.21.

Theorem 4.21. The Kalman-Bucy estimator of $X(t)$ is $\hat{X}(t)$ if and only if $\hat{X}(t)$ is the solution of (4.172), where $P(t)$ is the solution of (4.171).

Since $A(t)$, $B(t)$, $H(t)$ and $B^*(t)$ are assumed to be known, $P(t)$ may be solved numerically prior to the onset of an experiment of a filtering process, thus requiring no on-line computational work. As the experiment evolves, realizations of $\hat{X}(t)$ may be solved numerically and recursively (hence in real time) with the aid of (4.172) when $\mathbf{Z}(s)$ is replaced by the measured data $\mathbf{Z}(s, \omega)$, $s \in [0, T]$ (Sect. 4.1).

In the above procedure, $B^*(u)$ must be invertible at each $u \in [0, T]$. However, the importance of this requirement has been noted earlier in Sect. 4.5 in another context.

4.7.2 Supplementary Exercise

Exercise 4.8. Show that in (4.151) the order of integration may be interchanged.

5. A Theorem by Liptser and Shiryayev

A relatively complete mathematical development of the Kalman-Bucy filter has been presented in Chap. 4. One of the results that emerged in this discourse is the fact that success of the filter depends on the contamination of observation by white noise, much like dimming the light to see more clearly. The purpose of this short chapter is to shed some light on this self-contradictory phenomenon through a theorem by Liptser and Shiryayev.

5.1 Discussion on Observation Noise

We have seen that for second-order systems, linear minimum variance estimators are orthogonal projections onto closed subspaces of a Hilbert space. Hence, their existence and uniqueness are guaranteed by the projection theorem. However, methods for calculating orthogonal projections are available only in special cases. Apart from orthogonal projections in finite-dimensional spaces or in spaces with orthonormal bases (Fourier analysis), computational methods have been developed for estimates in wide sense stationary systems and for estimates of the Kalman-Bucy type. In both latter cases, the techniques are based on the possibility of representing the estimates as a unique and well-defined integral. In the first case, it is accomplished by virtue of the spectral representation of stationary processes [5.1] and, in the second, as a result of the corruption of observations by white noise, giving the integral

$$\int_0^t f^T(s)\, dZ(s)$$

where the components of $f(s)$ belong to $L_2[0, t]$ (Theorem 4.11).

Hence, in Kalman-Bucy filtering, the observation noise is not only a physical reality but also an essential tool in the computations. The computations carried out in Chap. 4 would break down in the absence of this noise. For noise-free observations, white noise may be added to the observation model to make the computations feasible and still obtain a "good" approximation of the estimator. What is discussed here will be made precise below.

5.2 A Theorem of Liptser and Shiryayev

Let $I = [0, T] \subset \mathbb{R}$, $X \in L_2(\Omega)$, $Y : I \to L_2(\Omega)$ and let $W : I \to L_2(\Omega)$ be a Wiener-Lévy process, orthogonal to X and Y, i.e., with s and t in I,

$$\begin{aligned} &\mathrm{E}W(s) = 0\,, \quad &&\mathrm{E}\{W(s)W(t)\} = \min(s, t) \\ &\mathrm{E}\{XW(s)\} = 0\,, \quad &&\mathrm{E}\{Y(s)W(t)\} = 0\,. \end{aligned}$$

Define

$$Z_n(s) = Y(s) + \frac{1}{n}W(s)\,, \qquad s \in I\,, \qquad n \in \mathbb{N} \tag{5.1}$$

and let $[Y]$, $[Z_n]$, $[W]$, $[Y, Z_n]$ and $[Y, W]$ be the closed linear hulls in $L_2(\Omega)$ of the sets $\{Y(s), s \in I\}$, $\{Z_n(s), s \in I\}$, $\{W(s), s \in I\}$, $\{Y(s); Z_n(t), s, t \in I\}$ and $\{Y(s); W(t), s, t \in I\}$, respectively. It is seen from (5.1) that

$$[Y, Z_n] = [Y, W]\,.$$

We also have

$$[W] \perp [Y]\,.$$

Let

$$\mathcal{P} : L_2(\Omega) \to [Y]$$

be the orthogonal projection onto $[Y]$ and let

$$\mathcal{P}_n : L_2(\Omega) \to [Z_n]$$

be the orthogonal projection onto $[Z_n]$ (Fig. 5.1). We shall state and prove the following result which is a part of a theorem by *Liptser* and *Shiryayev* [Ref. 5.2, Part I, p. 379]. This result formalizes the statement advanced in the introduction.

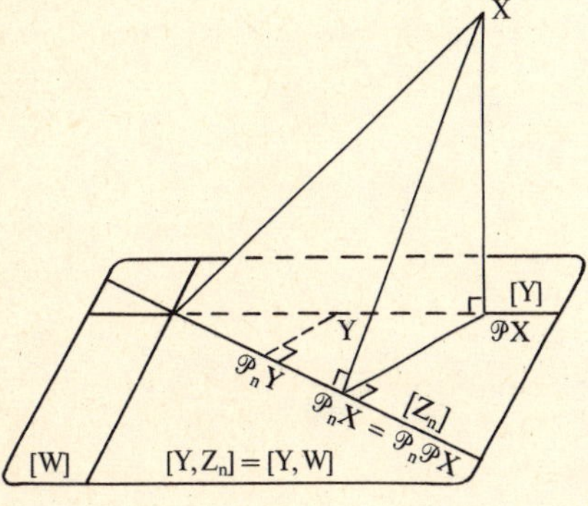

Fig. 5.1. Linear hulls and orthogonal projections

Theorem 5.1.

$$\forall X \in L_2(\Omega) : \lim_{n\to\infty} \mathcal{P}_n X = \mathcal{P} X \quad \text{in m.s.} \tag{5.2}$$

Proof. We shall first show that, for any $Y \in [Y]$,

$$\|Y - \mathcal{P}_n Y\| \to 0 \tag{5.3}$$

as $n \to \infty$. Consider $Y(s)$ at a fixed $s \in I$. Since $Z_n(s) \in [Z_n]$ and using (5.1), we have

$$\begin{aligned} \|Y(s) - \mathcal{P}_n Y(s)\|^2 &= \min_{V \in [Z_n]} \|Y(s) - V\|^2 \\ &\leq \|Y(s) - Z_n(s)\|^2 \\ &= \left\| \frac{1}{n} W(s) \right\|^2 = \frac{s}{n^2} \to 0 , \quad n \to \infty . \end{aligned} \tag{5.4}$$

If $Y \in [Y]$, it is the m.s. limit of some sequence $\{Y_m\}_{m \in \mathbb{N}}$ whose elements are linear combinations of the values $Y(s)$, $s \in I$, say

$$Y_m = y_{1m} Y(s_{1m}) + \ldots + y_{k_m m} Y(s_{k_m m})$$

$$y_{1m}, \ldots, y_{k_m m} \text{ in } \mathbb{R} , \quad s_{1m}, \ldots, s_{k_m m} \text{ in } I .$$

According to (5.4) and since $0 \leq s_{im} \leq T$, we now have

$$\begin{aligned} \|Y_m - \mathcal{P}_n Y_m\| &= \left\| \sum_{i=1}^{k_m} y_{im} Y(s_{im}) - \mathcal{P}_n \sum_{i=1}^{k_m} y_{im} Y(s_{im}) \right\| \\ &\leq \sum_{i=1}^{k_m} |y_{im}| \, \|Y(s_{im}) - \mathcal{P}_n Y(s_{im})\| \\ &\leq \left(\sum_{i=1}^{k_m} |y_{im}| \right) \frac{\sqrt{T}}{n} \to 0 \end{aligned} \tag{5.5}$$

as $n \to \infty$.

Finally, let

$$Y = \lim_{m\to\infty} Y_m \quad \text{in m.s.,}$$

where Y_m is of the type defined above. Since

$$\| \mathcal{P}_n (Y_m - Y)\| \leq \|Y_m - Y\| ,$$

then

$$\begin{aligned} \|Y - \mathcal{P}_n Y\| &\leq \|Y - Y_m\| + \|Y_m - \mathcal{P}_n Y_m\| + \| \mathcal{P}_n Y_m - \mathcal{P}_n Y\| \\ &\leq 2 \|Y - Y_m\| + \|Y_m - \mathcal{P}_n Y_m\| . \end{aligned} \tag{5.6}$$

With m fixed such that

$$\|Y - Y_m\| < \varepsilon , \quad \varepsilon > 0 ,$$

it is seen from (5.5) that there is a number $N \in \mathbb{N}$ such that $n > N$ implies

$$\|Y_m - \mathcal{P}_n Y_m\| < \varepsilon .$$

Hence, (5.6) gives

$$\|Y - \mathcal{P}_n Y\| < 3\varepsilon$$

if $n > N$, thus proving (5.3).

Therefore, since $\mathcal{P}X$ is an element of $[Y]$, (5.3) gives

$$\mathcal{P}X - \mathcal{P}_n \mathcal{P}X \to 0 \quad \text{as} \quad n \to \infty . \tag{5.7}$$

Let us next show

$$\mathcal{P}_n X = \mathcal{P}_n \mathcal{P}X . \tag{5.8}$$

This means that the orthogonal projection of X onto $[Z_n]$ coincides with that of $\mathcal{P}X$, or,

$$X - \mathcal{P}_n \mathcal{P}X \perp [Z_n] . \tag{5.9}$$

To prove (5.9), we write

$$X - \mathcal{P}_n \mathcal{P}X = (X - \mathcal{P}X) + (\mathcal{P}X - \mathcal{P}_n \mathcal{P}X) . \tag{5.10}$$

Here $X - \mathcal{P}X$ is orthogonal to $[Y]$ according to the definition of $\mathcal{P}$, and also to $[W]$ since $[W]$ is orthogonal to $[Y]$ and X. So $X - \mathcal{P}X$ is orthogonal to $[Y, W] = [Y, Z_n]$ and, since $[Z_n] \subset [Y, Z_n]$,

$$X - \mathcal{P}X \perp [Z_n] . \tag{5.11}$$

From the definition of $\mathcal{P}_n$,

$$\mathcal{P}X - \mathcal{P}_n \mathcal{P}X \perp [Z_n] . \tag{5.12}$$

Now (5.10–12) give (5.9, 8), whereas (5.8, 7) give (5.2), completing the proof. □

Appendix: Solutions to Selected Exercises

Exercise 2.1. According to Theorems 2.3, 4,

$$\frac{f(t+h)X(t+h)-f(t)X(t)}{h}$$

$$= \left[\frac{f(t+h)-f(t)}{h}\right]X(t+h) + f(t)\left[\frac{X(t+h)-X(t)}{h}\right]$$

$$\to f'(t)X(t) + f(t)X'(t) .$$

Exercise 2.2. a) X and $\mathrm{E}\{X(s)X(t)\}$ are presented graphically below.

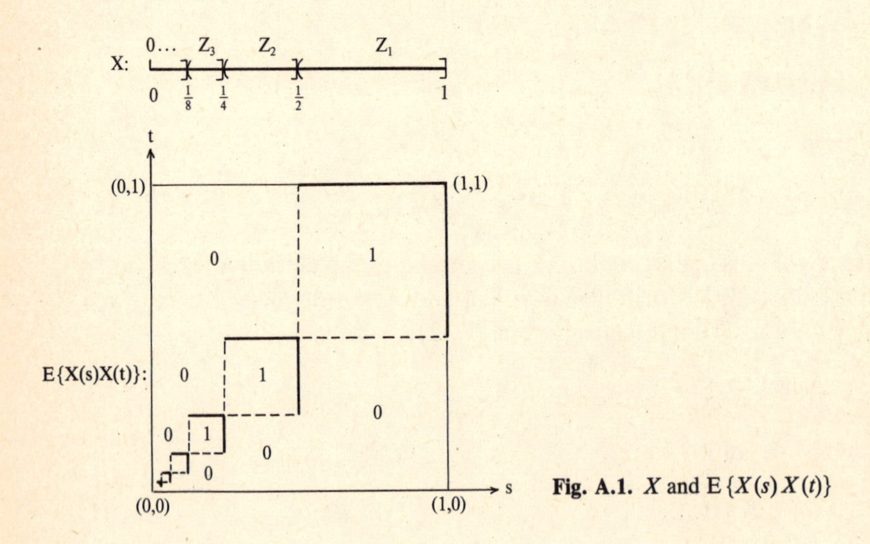

Fig. A.1. X and $\mathrm{E}\{X(s)X(t)\}$

b) It follows directly from (a).
c) If $h = k$,

$$\mathop{\Delta}_{h}\mathop{\Delta}_{h}\frac{\mathrm{E}\{X(0)X(0)\}}{h^2} = \frac{1}{h^2}\mathrm{E}\{X(h)-X(0)\}^2 = \frac{1}{h^2}\mathrm{E}X^2(h) = \frac{1}{h^2}$$

which does not converge as $h \to 0$.

Exercise 2.3. Let us show only that

$$\int_0^1 f(t)\,dX(t)$$

does not exist. Consider a convergent sequence $\{p_n\}_{n\in N}$ of partitions of [0, 1] such that in each p_n the intermediate points do not coincide with the subdivision points and such that the point 1/2 is a subdivision point of p_n if n is odd, but not if n is even. We can show that, in this case,

$$S_{f,X}(p_n) = \begin{cases} (1)\,(0-Z) = -Z & \text{if } n \text{ is odd} \\ 0 & \text{if } n \text{ is even.} \end{cases}$$

It follows that

$$\{S_{f,X}(p_n)\}_{n\in N} = -Z, 0, -Z, 0, -Z, 0, \ldots .$$

Hence, it is not a Cauchy sequence in $L_2(\Omega)$.

Exercise 2.4. It is seen that for each partition p_n of [0, 1],

$$S_{f,X}(p_n) = \sum_i c\{X(t_i) - X(t_{i-1})\} = c\{X(t) - X(0)\} .$$

Hence, the integral

$$\int_0^1 f(t)\,dX(t)$$

exists and

$$\int_0^1 f(t)\,dX(t) = c\{X(t) - X(0)\} .$$

Regarding

$$\int_0^1 X(t)\,df(t) ,$$

all Riemann-Stieltjes sums equal zero. Hence, the integral exists and

$$\int_0^1 X(t)\,df(t) = 0 .$$

Exercise 2.6. All Riemann-Stieltjes sums corresponding to

$$\int_0^1 f(t)\,dX(t)$$

are zero and those corresponding to

$$\int_1^2 f(t)\,dX(t)$$

equal C. Hence, these integrals exist (and are equal to zero and C, respectively).

For

$$\int_0^2 f(t)\,dX(t)$$

let $\{p_n\}_{n\in N}$ be a convergent sequence of partitions of [0, 2] such that none of p_n contains 1 as a subdivision point and such that the intermediate point in the partition interval containing 1 is chosen to be left of 1 if n is odd, and right of 1 if n is even. Then,

$$S_{f,X}(p_n) = \begin{cases} 0 & \text{if } n \text{ is odd} \\ C & \text{if } n \text{ is even} \end{cases} .$$

Hence, $\{S_{f,X}(p_n)\}_{n\in N}$ does not converge and the integral

$$\int_0^2 f(t)\,dX(t)$$

does not exist.

Exercise 2.8. Let p be a partition of I with subdivision points $t_o, t_1, \ldots, t_n$ such that

$$a = t_o < t_1 \ldots < t_n = b .$$

Set

$$\Delta_i X = X(t_i) - X(t_{i-1}) , \quad i = 1, \ldots, n .$$

Then

$$\Delta_i \mathrm{E}X = \mathrm{E}\{\Delta_i X\}$$

and, using the second inequality of (1.67),

$$V_{\mathrm{E}X}(p) = \sum_{i=1}^n |\Delta_i \mathrm{E}X| \leqslant \sum_{i=1}^n \|\Delta_i X\| = V_X(p)$$

and we have the assertion.

Exercise 2.11. Let $t_o, t_1, \ldots, t_n$ be the subdivision points of a partition p of I such that

$$a = t_o < t_1 < \ldots < t_n = b .$$

Set

$$C_X(s, t) = \mathrm{E}\{X(s)X(t)\} , \quad (s, t) \in I^2$$

$$m(t) = \mathrm{E}X(t)$$

$$X(t) = m(t) + Y(t) \quad \text{entailing} \quad \mathrm{E}Y(t) = 0 , \quad t \in I$$

$$\begin{aligned}
\Delta_i X &= X(t_i) - X(t_{i-1}) \\
\Delta_i m &= m(t_i) - m(t_{i-1}) \\
\Delta_i Y &= Y(t_i) - Y(t_{i-1}), \quad i = 1, \ldots, n .
\end{aligned}$$

Then

$$\Delta_i X = \Delta_i m + \Delta_i Y , \quad \text{and if}$$

$$r_{ij} = [t_{i-1}, t_i] \times [t_{j-1}, t_j] , \quad \text{then}$$

$$\begin{aligned}
\underset{r_{ij}}{\Delta\Delta} C_x &= \mathrm{E}\{\Delta_i X \Delta_j X\} = \mathrm{E}\{(\Delta_i m + \Delta_i Y)(\Delta_j m + \Delta_j Y)\} \\
&= \Delta_i m \Delta_j m + \mathrm{E}\{\Delta_i Y \Delta_j Y\} \text{ (since } \mathrm{E}Y(t) = 0).
\end{aligned}$$

Now we define the set of numbers $e_1, \ldots, e_n$ as follows:

$$e_i = \begin{cases} 1 & \text{if } \Delta_i m \geqslant 0 \\ -1 & \text{if } \Delta_i m < 0 . \end{cases}$$

Then,

$$\begin{aligned}
V_{C_x}(p \times p) &= \sum_{i=1}^{n} \sum_{j=1}^{n} |\mathrm{E}\{\Delta_i X \Delta_j X\}| \geqslant \sum_{i=1}^{n} \sum_{j=1}^{n} (\mathrm{E}\{\Delta_i X \Delta_j X\})\, e_i e_j \\
&= \sum_{i=1}^{n} \sum_{j=1}^{n} (\Delta_i m \Delta_j m\, e_i e_j) + \sum_{i=1}^{n} \sum_{j=1}^{n} (\mathrm{E}\{\Delta_i Y \Delta_j Y\})\, e_i e_j \\
&= \left(\sum_{i=1}^{n} |\Delta_i m| \right)^2 + \mathrm{E}\left\{ \sum_{i=1}^{n} (\Delta_i Y)\, e_i \right\}^2 \geqslant \left(\sum_{i=1}^{n} |\Delta_i m| \right)^2 \\
&= \{V_m(p)\}^2 .
\end{aligned}$$

The assertion follows by taking limits if p runs through a convergent sequence of partitions of I.

Exercise 2.12. We have, formally,

$$\begin{aligned}
&\mathrm{E}\left\{ \int_0^s F(u)\, dW(u) \left[\int_0^t G(v)\, dW(v) \right]^T \right\} \\
&= E\left\{ \int_0^s F(u)\, \dot{W}(u)\, du \int_0^t \dot{W}^T(v)\, G^T(v)\, dv \right\} \\
&= \int_0^s du \int_0^t dv\, F(u)\, \mathrm{E}\{\dot{W}(u)\, \dot{W}^T(v)\}\, G^T(v) \\
&= \int_0^t \left[\int_0^s F(u)\, B(u)\, \delta(u - v)\, du \right] G^T(v)\, dv \\
&= \int_0^s F(v)\, B(v)\, G^T(v)\, dv
\end{aligned}$$

if $0 \leqslant s \leqslant t \leqslant T$.

Exercise 2.14. Define

$$X(\omega, t) = \sqrt{\omega + t}, \quad (\omega, t) \in [0, 1] \times [0, T].$$

Then the sample derivative is

$$X'(\omega, t) = \frac{1}{2\sqrt{\omega + t}}$$

in the given domain, outside the point (0, 0). Hence, the samples are differentiable on $[0, T]$ with probability one.

Now, X is of second order since

$$\mathrm{E}X^2 = \int_0^1 (\omega + t)\, d\omega < \infty$$

and it is m.s. differentiable if $t > 0$ with derivative $1/(2\sqrt{\omega + t})$. However, $X'(\omega, t)$ is not of second order at $t = 0$ since

$$\mathrm{E}\{X'(\omega, 0)\}^2 = \int_0^1 \frac{1}{4\omega}\, d\omega$$

is not finite. It thus follows from Theorem 2.36 that X is not m.s. differentiable at $t = 0$.

Exercise 3.1. Let $X_i \in L_2(\Omega)$ be the components of X and $m_{ij} \in \mathbb{R}$ the elements of M. $(X, Y)_N$ is an inner product in $L_2^N(\Omega)$ because the following conditions hold for all X, Y, Z in $L_2^N(\Omega)$ and for all $\alpha \in \mathbb{R}$:

a) since M is symmetric,

$$(X, Y)_N = \mathrm{E}\{X^T M Y\} = \mathrm{E}\{Y^T M X\} = (Y, X)_N$$

b) $(\alpha X, Y)_N = \alpha (X, Y)_N$

c) $(X + Y, Z)_N = (X, Z)_N + (Y, Z)_N$

d) the symmetry of M implies that there is an orthogonal matrix U such that

$$M = U^T \operatorname{diag}(\lambda_1, \ldots, \lambda_N)\, U.$$

The eigenvalues λ_i of M are real and positive since M is positive. Hence, with u_{ij} denoting the elements of U,

$$\begin{aligned}(X, X)_N &= \mathrm{E}\{X^T M X\} = \mathrm{E}\{X^T U^T \operatorname{diag}(\lambda_1, \ldots, \lambda_N)\, U X\} \\ &= \sum_{i=1}^{N} \lambda_i \, \mathrm{E}\left\{\sum_{j=1}^{N} u_{ij} X_j\right\}^2 > 0\end{aligned}$$

unless

$$\sum_{j=1}^{N} u_{ij} X_j = 0$$

for all i, i.e.,

$$UX = 0\,, \quad X = 0 \text{ a.s.}$$

Exercise 3.3. According to (3.57), we have formally

$$\begin{aligned} X(t) &= \Phi(t)\left\{C + \int_0^t \Phi^{-1}(s)\,\dot{W}(s)\,ds\right\} \\ &= \Phi(t)\left\{C + \int_0^t \Phi^{-1}(s)\frac{d}{ds}W(s)\,ds\right\} \\ &= \Phi(t)\left\{C + \int_0^t \Phi^{-1}(s)\,dW(s)\right\}. \end{aligned}$$

Exercise 4.1. If U, V and W are subspaces of $L_2(\Omega)$ such that $U \perp V$ and $W = U + V (= U \oplus V)$, and if $\mathcal{P}_U$, $\mathcal{P}_V$ and $\mathcal{P}_W$ are the orthogonal projection operators of $L_2(\Omega)$ onto U, V and W, respectively, then

$$\mathcal{P}_W = \mathcal{P}_U + \mathcal{P}_V\,,$$

i.e., for any $X \in L_2(\Omega)$, $\mathcal{P}_W X = \mathcal{P}_U X + \mathcal{P}_V X$. By definition, $X - \mathcal{P}_U X \in U^\perp$, $\mathcal{P}_V X \in V \subset U^\perp$, and hence $X - \mathcal{P}_U X - \mathcal{P}_V X \in U^\perp$, or

$$X - (\mathcal{P}_U X + \mathcal{P}_V X) \perp U\,.$$

Analogously,

$$X - (\mathcal{P}_U X + \mathcal{P}_V X) \perp V \quad \text{and so}$$

$$X - (\mathcal{P}_U X + \mathcal{P}_V X) \perp U + V = W\,.$$

As also

$$\mathcal{P}_U X + \mathcal{P}_V X \in U + V = W\,,$$

we have

$$\mathcal{P}_W X = \mathcal{P}_U X + \mathcal{P}_V X\,.$$

Applying this result to $[Z] = [D] \oplus [Z_o]$ and with $X = \mathrm{E}X + X_o$, we obtain

$$\begin{aligned} \mathcal{P}_Z X &= \mathcal{P}_Z \mathrm{E}X + \mathcal{P}_Z X_o \\ &= \mathcal{P}_D \mathrm{E}X + \mathcal{P}_{Z_o} \mathrm{E}X + \mathcal{P}_D X_o + \mathcal{P}_{Z_o} X_o\,, \end{aligned} \tag{A.1}$$

where $\mathrm{E}X$, looked upon as a degenerate random variable, is an element of $[D]$. Hence,

$$\mathcal{P}_D \mathrm{E}X = \mathrm{E}X\,.$$

Also, since $[D] \perp [Z_o]$,

$$\mathcal{P}_{Z_o} \mathrm{E}X = 0 .$$

As $X_o \perp [D]$,

$$\mathcal{P}_D X_o = 0$$

and so finally (A.1) gives

$$\mathcal{P}_Z X = \mathrm{E}X + \mathcal{P}_{Z_o} X_o .$$

Exercise 4.5. Let $V \subset L_2(\Omega)$ be a finite-dimensional subspace, let $\phi_1, \ldots, \phi_N$ be an orthonormal base of V and let

$$\{f_n\}_{n \in \mathbb{N}}$$

be a Cauchy sequence in V. For each $n \in \mathbb{N}$, there are real numbers $c_{n1}, \ldots, c_{nN}$ such that

$$f_n = c_{n1} \phi_1 + \ldots + c_{nN} \phi_N .$$

It is seen that

$$\|f_m - f_n\|^2 = (c_{m1} - c_{n1})^2 + \ldots + (c_{mN} - c_{nN})^2$$

and

$$\{c_{n1}\}_{n \in \mathbb{N}}, \ldots, \{c_{nN}\}_{n \in \mathbb{N}}$$

are Cauchy sequences in $\mathbb{R}$, hence convergent with limits, say $c_1, \ldots, c_N$, respectively.

It follows that f_n converges to $c_1\phi_1 + \ldots + c_N\phi_N \in V$ as $n \to \infty$. Hence V is closed.

Exercise 4.6. The mapping $\mathcal{F} : L_2([0, t]) \to H(W, t)$ is an isometry since, for all f and g in $L_2([0, t])$,

$$\left(\int_0^t f(u)\, dW(u), \int_0^t g(v)\, dW(v) \right)_\Omega$$

$$= \mathrm{E}\left\{ \int_0^t f(u)\, dW(u) \int_0^t g(v)\, dW(v) \right\}$$

$$= \int_0^t f(s)\, g(s)\, ds = (f, g)_{[0, t]}$$

due to Theorem 4.9, where $b_{ij} = 1$ since W_i and W_j are identical to the standard Wiener-Lévy process.

Exercise 4.7. For all f in $L_2([0, t])$, see (4.53),

$$\mathscr{F}f = \int_0^t f(s)\,dZ(s) = \int_0^t f(s)\,R(s)\,ds + \int_0^t f(s)\,dW(s) \in H(Z, t)\,.$$

Clearly, $\mathscr{F}: L_2([0, t]) \to H(Z, t)$ is linear. It is shown in D of 4.5 that $\mathscr{F}$ is surjective and 1–1. Hence,

$$\mathscr{F}^{-1}: H(Z, t) \to L_2([0, t])$$

is defined (and linear since $\mathscr{F}$ is linear).

We shall show that both $\mathscr{F}$ and $\mathscr{F}^{-1}$ are bounded. Set $X = \mathscr{F}f, f = \mathscr{F}^{-1}X$. Because of the orthogonality of $R(u)$ and $W(v)$ we have

$$\|X\|_\Omega^2 = \|\mathscr{F}f\|_\Omega^2 = \left\| \int_0^t f(s)\,R(s)\,ds + \int_0^t f(s)\,dW(s) \right\|_\Omega^2$$

$$= \left\| \int_0^t f(s)\,R(s)\,ds \right\|_\Omega^2 + \left\| \int_0^t f(s)\,dW(s) \right\|_\Omega^2 . \tag{A.2}$$

From (A.2) it follows that

$$\left\| \int_0^t f(s)\,dW(s) \right\|_\Omega^2 \leq \|X\|_\Omega^2 ,$$

i.e.,

$$\|f\|_{[0,t]}^2 = \int_0^t f^2(s)\,ds \leq \|X\|_\Omega^2$$

$$\|\mathscr{F}^{-1}X\|_{[0,t]}^2 \leq (1)\,\|X\|_\Omega^2 \tag{A.3}$$

and

$$\|\mathscr{F}f\|^2 = \int_0^t \int_0^t f(u) f(v)\,\mathrm{E}\{R(u)\,R(v)\}\,du\,dv + \int_0^t f^2(s)\,ds\,.$$

And so, if $M = \max \mathrm{E}\{R(u)\,R(v)\}$ on $[0, t]^2$,

$$\|\mathscr{F}f\|_\Omega^2 \leq M\left[\int_0^t f(s)\,ds\right]^2 + \int_0^t f^2(s)\,ds$$

$$\|\mathscr{F}f\|_\Omega^2 \leq Mt \int_0^t f^2(s)\,ds + \int_0^t f^2(s)\,ds$$

$$\|\mathscr{F}f\|_\Omega^2 \leq (Mt + 1)\,\|f\|_{[0,t]}^2 \tag{A.4}$$

then (A.3, 4) show the boundedness of $\mathscr{F}^{-1}$ and $\mathscr{F}$, respectively.

References

Chapter 1

1.1 M. Loéve: *Probability Theory* (Van Nostrand, New York 1963)
1.2 K. L. Chung: *A Course in Probability Theory* (Harcourt, Brace and World, New York 1968)
1.3 T. T. Soong: *Probabilistic Modeling and Analysis in Science and Engineering* (Wiley, New York 1981)
1.4 B. R. Frieden: *Probability, Statistical Optics and Data Testing,* Springer Ser. Inf. Sci., Vol. 10 (Springer, Berlin, Heidelberg, 1983)
1.5 J. L. Doob: *Stochastic Processes* (Wiley, New York 1953)

Chapter 2

2.1 M. Loéve: *Probability Theory* (Van Nostrand, New York 1963)
2.2 T. T. Soong: *Random Differential Equations in Science and Engineering* (Academic, New York 1973)

Chapter 3

3.1 E. Hille, R. S. Phillips: *Functional Analysis and Semigroups* (American Mathematical Society, Providence, RI 1957)
3.2 E. Hille: *Lectures on Ordinary Differential Equations* (Addison-Wesley, Reading, MA 1969)
3.3 E. Wong, M. Zakai: *On the Convergence of Ordinary Integrals to Stochastic Integrals*, Ann. Math. Stat. **36,** 1560–1564 (1965)
3.4 P. A. Ruymgaart, T. T. Soong: *A Sample Treatment of Langevin-Type Stochastic Differential Equations*, J. Math. Analysis Appl. **34,** 325–338 (1971)
3.5 A. H. Zemanian: *Distribution Theory and Transform Analysis* (McGraw-Hill, New York 1965)

Chapter 4

4.1 R. E. Kalman, R. S. Bucy: *New Results in Linear Filtering and Prediction Theory*, Trans. ASME J. Basic Eng. **83,** 95–107 (1961)
4.2 E. Hewitt, K. Stromberg: *Real and Abstract Analysis. A Modern Treatment of the Theory of Functions of a Real Variable*. Graduate Texts in Mathematics, Vol. 25 (Springer, Berlin, Heidelberg, New York 1975)
4.3 T. Kailath: *An Innovations Approach to Least-square Estimation – Part I: Linear Filtering in Additive White Noise, IEEE Trans.* **AC-13,** 646–655 (1968)
4.4 A. E. Taylor: *Introduction to Functional Analysis* (Wiley, New York 1958)
4.5 N. I. Achieser, I. M. Glasmann: *Theorie der Linearen Operatoren in Hilbert-Raum* (Akademie, Berlin 1968)
4.6 P. A. Ruymgaart: *A Note on the Integral Representation of the Kalman-Bucy Estimate*, Indag. Math. **33,** 346–360 (1971)

4.7 P. A. Ruymgaart: *The Kalman-Bucy Filter and its Behavior with respect to Smooth Perturbations of the Involved Wiener-Lévy Processes*, Ph.D. Dissertation, Technological University Delft, Delft (1971)
4.8 R. S. Bucy, P. D. Joseph: *Filtering for Stochastic Processes with Applications to Guidance* (Wiley-Interscience, New York 1968)
4.9 K. Karhunen: *Über Lineare Methoden in der Wahrscheinlichkeitsrechnung* Ann. Ac. Fennicae, Ser. A.1, Mathematica-Physica, **73** (1947)
4.10 T. S. Huang (ed.): *Two-Dimensional Digital Signal Processing I*, Topics in Appl. Phys., Vol. 42 (Springer, Berlin, Heidelberg, New York 1981)

Chapter 5

5.1 M. Loève: *Probability Theory* (Van Nostrand, Princeton 1955)
5.2 R. S. Liptser, A. N. Shiryayev: *Statistics of Random Process I, General Theory,* Applications of Mathematics, Vol. 5 (Springer, Berlin, Heidelberg, New York 1978)

Subject Index

Almost sure (a.s.) 4
Almost everywhere (a.e.) 104

Banach space 82, 140
Borel algebra 2
Borel function 2
Bounded variation 46, 55
  in strong sense 52
  in the weak sense 57
Brownian motion 27

Calculus in m.s. 30
Cauchy sequence 23, 30, 82, 104
Cauchy's inequality 22, 23, 31, 104, 137
Centered stochastic process 38
Characteristic function 9
Chebyshev's inequality 24
Complete space 24, 31, 82, 105
Continuity in m.s. 33
Convergence in m.s. 30
Correlation function 25
Correlation (function) matrix 25
Covariance function 25
Covariance (function) matrix 26
Cross correlation function 25
Cross covariance function 25
Cross covariance (function) matrix 26

Degenerate r.v. 7
Degenerate stochastic process 38
Differentiability in m.s. 36
Distance 23, 104
Dynamic system 80, 92, 146

Error covariance matrix 149
Event 1

Filtering 80
Fundamental sequence 23
Fundamental solution 89

Gaussian distribution 16
Gaussian manifold 17
Gaussian-Markov vector process 93
Gaussian *N*-vector 16
Gaussian process 26
Gaussian r.v. 10
General Wiener-Lévy *N*-vector 72

Hilbert space 24, 105, 136

Increment 28
Independence (stochastic) 13–15
Inner product 22, 104, 136
Intermediate point 40

Joint characteristic function 13
Joint probability density function 12
Joint probability distribution function 11

Kailath's innovations approach 116
Kalman-Bucy estimator 101, 116, 128
Kalman-Bucy filtering 80, 100–102, 128, 154
Kalman filter 100

Least squares 101, 114
Lebesgue measure 3
Lebesgue measure space 3
Limit in mean (square) (l.i.m.) 31
Linear hull 27
Linear least-squares estimation 101
Linear minimum variance estimator 101
Liptser and Shiryayev, theorem of VI, 155

Marginal distribution function 12
Markov vector process 93
Mathematical expectation 5
Mean 5
Mean square (m.s.) 30
Measurable set 2
Measurable space 2
Measure 2
  space 2

Noise 28, 75, 80
Norm 23, 81, 104, 136, 137, 140
Normal *N*-vector 16
Normal r.v. 10

Observation 100, 154
  noise VI, 100, 154
  process 100
Orthogonal 15, 22
  projection 101, 114, 115, 128, 154, 155

Partial integration 42
Partition 40, 54
Probability 1, 3
  density function 8
  distribution function 7
  measure 3
  space 3

Random experiment 1
Random process 18
Random variable (r.v.) 4
Random vector 11
Realization 18
Reconstruction 102
Refinement 41, 54
Riccati differential equation 152
Riemann integral in m.s. 42
Riemann sum 42
Riemann-Stieltjes (R-S) integral 56
R-S integral in m.s. 41
R-S sum 41, 56

Sample calculus 76
Sample function 18
Sample solution 81
Sample space 1
Second order process 24
Second order r.v. 21
Simple r.v. 4
Solution in m.s. sense 81
Standard Wiener-Lévy *N*-vector 29, 70
Standard Wiener-Lévy process 28, 68
State vector 80
Stochastic dynamic system 80, 81
Stochastic independence 13, 14, 15
Stochastic process 18
Subdivision point 40

Total variation 46, 52, 55
Trajectory 18
Tychonoff, theorem of 124

Uniformly m.s. continuity 34

Variance 7
Variation 46, 52, 55

White noise 28, 75, 80
Wiener-Hopf equation 130
Wiener-Lévy *N*-vector 29, 72
Wiener-Lévy process 28, 68
Wong and Zakai, theorem of 99

Digital Pattern Recognition

Editor: **K.S.Fu**
With contributions by numerous experts
2nd corrected and updated edition. 1980. 59 figures, 7 tables. XI, 234 pages. (Communication and Cybernetics, Volume 10). ISBN 3-540-10207-8

Contents: Introduction. – Topics in Statistical Pattern Recognition. – Clustering Analysis. – Syntactic (Linguistic) Pattern Recognition. – Picture Recognition. – Speech Recognition and Understanding. – Recent Developments in Digital Pattern Recognition. – Subject Index.

Syntactic Pattern Recognition, Applications

Editor: **K.S.Fu**
1977. 135 figures, 19 tables. XI, 270 pages. (Communication and Cybernetics, Volume 14). ISBN 3-540-07841-X

Contents: *K.S.Fu:* Introduction to Syntactic Pattern Recognition. – *S.Horowitz:* Peak Recognition in Waveforms. – *J.E.Albus:* Electrocardiogram Interpretation Using a Stochastic Finite-State Model. – *R.DeMori:* Syntactic Recognition of Speech Patterns. – *W.W.Stallings:* Chinese Character Recognition. – *T.Pavlidis, H.-Y.F.Feng:* Shape Discrimination. – *R.H.Anderson:* Two-Dimensional Mathematical Notation. – *B.Moayer, K.S.Fu:* Fingerprint Classification. – *J.M.Brayer, P.H.Swain, K.S.Fu:* Modeling of Earth Resources Satellite Data. – *T.Vámos:* Industrial Objects and Machine Parts Recognition.

Springer-Verlag
Berlin
Heidelberg
New York
Tokyo

Computer Processing of Electron Microscope Images

Editor: **P.W.Hawkes**
1980. 116 figures, 2 tables. XIV, 296 pages. (Topics in Current Physics, Volume 13). ISBN 3-540-09622-1

Contents: *P.W.Hawkes:* Image Processing Based on the Linear Theory of Image Formation. – *W.O.Saxton:* Recovery of Specimen Information for Strongly Scattering Objects. – *J.E.Mellema:* Computer Reconstruction of Regular Biological Objects. – *W.Hoppe, R.Hegerl:* Three-Dimensional Structure Determination by Electron Microscopy (Nonperiodic Specimens). – *J.Frank:* The Role of Correlation Techniques in Computer Image Processing. – *R.H.Wade:* Holographic Methods in Electron Microscopy. – *M.Isaacson, M.Utlaut, D.Kopf:* Analog Computer Processing of Scanning Transmission Electron Microscope Images.

Picture Processing and Digital Filtering

Editor: **T. S. Huang**

2nd corrected and updated edition. 1979.
113 figures, 7 tables. XIII, 297 pages
(Topics in Applied Physics, Volume 6)
ISBN 3-540-09339-7

Contents: *T. S. Huang:* Introduction. – *H. C. Andrews:* Two-Dimensional Transforms. – *J. G. Fiasconaro:* Two-Dimensional Nonrecursive Filters. – *R. R. Read, J. L. Shanks, S. Treitel:* Two-Dimensional Recursive Filtering. – *B. R. Frieden:* Image Enhancement and Restoration. – *F. C. Billingsley:* Noise Considerations in Digital Image Processing Hardware. – *T. S. Huang:* Recent Advances in Picture Processing and Digital Filtering. – Subject Index.

Two-Dimensional Digital Signal Processing I

Linear Filters

Editor: **T. S. Huang**

1981. 77 figures. X, 210 pages
(Topics in Applied Physics, Volume 42)
ISBN 3-540-10348-1

Contents: *T. S. Huang:* Introduction. – *R. M. Mersereau:* Two-Dimensional Nonrecursive Filter Design. – *P. A. Ramamoorthy, L. T. Bruton:* Design of Two-Dimensional Recursive Filters. – *B. T. O'Connor, T. S. Huang:* Stability of General Two-Dimensional Recursive Filters. – *J. W. Woods:* Two-Dimensional Kalman Filtering.

Two-Dimensional Digital Signal Processing II

Transforms and Median Filters

Editor: **T. S. Huang**

1981. 49 figures. X, 222 pages. (Topics in Applied Physics, Volume 43)
ISBN 3-540-10359-7

Contents: *T. S. Huang:* Introduction. – *J.-O. Eklundh:* Efficient Matrix Transposition. – *H. J. Nussbaumer:* Two-Dimensional Convolution and DFT Computation. – *S. Zohar:* Winograd's Discrete Fourier Transform Algorithm. – *B. I. Justusson:* Median Filtering: Statistical Properties. – *S. G. Tyan:* Median Filtering: Deterministic Properties.

Pattern Formation by Dynamic Systems and Pattern Recognition

Proceedings of the International Symposium on Synergetics at Schloß Elmau, Bavaria, April 30–May 5, 1979

Editor: **H. Haken**

1979. 156 figures, 16 tables. VIII, 305 pages
(Springer Series in Synergetics, Volume 5)
ISBN 3-540-09770-8

Contents: Introduction. – Temporal Patterns: Laser Oscillations and Other Quantum-Optical Effects. – Pattern Formation in Fluids. – Turbulence and Chaos. – Pattern Recognition and Pattern Formation in Biology. – Pattern Recognition and Associations. – Pattern Formation in Ecology, Sociology and History. – General Approaches. – Index of Contributors.

Springer-Verlag Berlin Heidelberg New York Tokyo